Clinical Enzymology

7th International Congress of Clinical Chemistry
Geneva (Switzerland)/Evian (France), September 8–13, 1969
General Editor: M. Roth, Geneva

Vol. 2

Clinical Enzymology

Editors: J. Frei and M. Jemelin, Lausanne

With 66 figures and 49 tables

19 70

University Park Press, Baltimore, London, Tokyo

Originalily published by S. Karger AG, Basel, Switzerland
Distributed exclusively in the United States of America and Canada
by University Park Press, Baltimore, Maryland

Library of Congress Catalog Card Number 79-145821
International Standard Book Number (ISBN) 0-8391-0590-8

7th International Congress of Clinical Chemistry
Geneva (Switzerland)/Evian (France), September 8–13, 1969
General Editor: M. Roth, Geneva

Vol. 1: Methods in Clinical Chemistry. Editor: M. ROTH, Geneva.
XIV+321 p., 142 fig., 49 tab., 1970.

Vol. 2: Clinical Enzymology. Editors: J. FREI and M. JEMELIN, Lausanne.
XIV+204 p., 66 fig., 49 tab., 1970.

Vol. 3: Hormones, Lipids and Miscellaneous. Editors: J. P. FELBER, Lausanne, and J.-J. SCHEIDEGGER, Geneva.
XIV+474 p., 235 fig., 71 tab., 1970.

Vol. 4: Digestion and Intestinal Absorption. Editors: P. HORE and G. SEMENZA, Zurich.
XVIII+134 p., 49 fig., 25 tab., 1970

S. Karger AG, Arnold-Böcklin-Strasse 25, CH-4000 Basel 11 (Switzerland)

Printed in Switzerland by Imprimerie Corbaz SA, Montreux
Blocks: Steiner & Co., Basel

Contents Vol. 1–4

Vol. 2: Clinical Enzymology

Symposium on Erythrocyte Enzymopathology

Symposium on Proteases Inhibitors

Communications

a) Methods of Enzyme Determination

b) Enzymes in Pathology

Vol. 1: Methods in Clinical Chemistry

Photometric Methods

Symposium on Fluorimetric Technics in Clinical Chemistry

Electroanalysis

Proteins

Drugs

Automation

Quality Control

Vol. 3: Hormones, Lipids and Miscellaneous

Hormones

Symposium on Radioimmunoassays

Methodology

Clinical Application of Radioimmunoassays

Lipids

Symposium. Problems of Detection of Metabolic Inborn Errors

Miscellaneous

Vol. 4: Digestion and Intestinal Absorption

Plenary Lectures

Symposium on Digestion and Intestinal Absorption

Contributed Paper

Foreword

The large part of this Congress devoted to enzymology shows once more the importance of this field in clinical medicine as well as in the area of experimental medicine. This volume contains the reports of two symposia and some free communications which concern either technical innovations for enzyme assays, or enzyme patterns in human pathology. One of the symposia was devoted to enzyme defects of erythrocytes, specifically hereditary defects and recent theories of aetiology. In the other symposium certain antiproteases were studied, primarily inhibitors of proteolytic enzymes involved in digestion or coagulation; the discussion centered around their determination, mode of action as well as their importance as metabolic regulators.

J. Frei.

7th int. Congr. clin. Chem., Geneva/Evian 1969; vol. 2: Clinical Enzymology, pp. 1–18
(Karger, Basel/München/Paris/New York 1970)

Inherited Abnormalities of Red Cell Glycolytic Enzymes

J. C. KAPLAN and C. KISSIN
Institute of Molecular Pathology, University Center Cochin, Paris,
and Laboratory of Enzymology, Children's Hospital Debrousse, Lyon

Fifteen years ago, when an intensive research program was undertaken to explain primaquine-induced anemia, there was no inherited enzyme deficiency known to occur in the red cells. Now in 1969, there are at least 15 different enzyme deficiencies which have been defined as responsible for disorders of the red cell. They can be categorized according to the particular metabolic pathway which is involved. Because the mature erythrocyte has a drastically simplified metabolism, a classification of these inherited enzyme defects can be easily made.

I. The first category concerns the glycolytic pathway, also known as the Embden-Meyerhof pathway.

II. The second category concerns the hexose monophosphate shunt.

III. The third category comprises miscellaneous enzymes involved in glutathione metabolism and methemoglobin reduction.

A fourth category would be those enzymes whose deficiency is detectable in the red cells, but does not alter their function, shape or survival.

Our purpose is to consider the enzyme abnormalities belonging to category I, i.e. the glycolytic pathway (fig. 1). We will include in this review the by-pass pathway located between 1,3-diphosphoglycerate (1,3-DPG) and 3-phosphoglycerate (3-PG), often referred to as the Rapoport-Luebering cycle, in which a high energy phosphate bond is lost with production of 2,3-diphosphoglycerate. (2,3-DPG).

At present, on the glycolytic pathway, 9 different enzyme deficiencies have been defined, among which 6 are well-documented, whereas 3 others still need confirmation (table I).

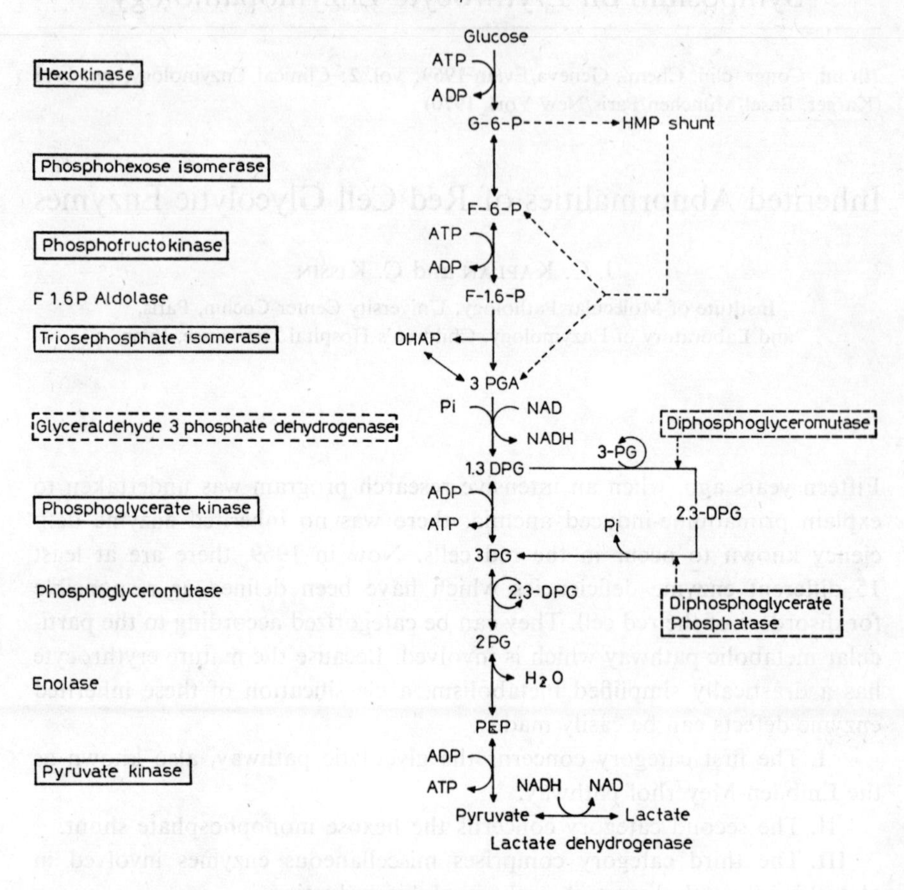

Fig. 1. Glycolysis in the red cell.

[solid box] Well-established enzyme deficiencies.

[dashed box] Presumed enzyme deficiencies, requiring further investigation.

As a preliminary point, it is interesting to emphasize that they all give rise to a hematological pattern of congenital non spherocytic hemolytic disease (CNHD), with a variable degree of severity.

Table I. Inherited abnormalities of red cell glygolytic enzymes

Enzyme deficiency	First report	Year	Incidence	Comments
Pyruvate kinase	VALENTINE *et al.*	1961	Not rare	Polymorphism suspected
Diphosphoglycero-mutase	PRANKERD	1961	Rare	Indirect evidence
Diphosphoglycerate phosphatase	JACOBASH *et al.*	1964	Rare	Needs confirmation
Triosephosphate isomerase	SCHNEIDER *et al.*	1965	Rare	Generalized disease
Phosphofructokinase	TARUI *et al.*	1965	Rare	A form of muscular glycogenosis (Type VII)
Glyceraldehyde 3-phosphate dehydrogenase	HARKNESS	1966	Rare	Needs confirmation
Hexokinase	VALENTINE *et al.*	1967	Rare	
Phosphohexose isomerase	BAUGHAN *et al.*	1968	Rare	
Phosphoglycerate kinase	VALENTINE *et al.*	1968	Rare	Sex linked trait

I. Enzymes of the Direct Embden-Meyerhof Pathway

A. Pyruvate Kinase

Pyruvate kinase (PK) deficiency was reported by VALENTINE *et al.* in 1961 [58, 65]. Since this first report, over 100 cases have appeared in the literature [60]. It now appears that pyruvate kinase deficiency ranks as the second most common cause of CNHD, behind glucose-6-phosphate dehydrogenase (G6PD) deficiency, but far ahead of the other enzyme deficiencies presently characterized. This recessive autosomal condition has been described in various countries and ethnic groups but seems to be more common in people of North European origin [59]. The clinical pattern is variable. The hemolytic syndrome can be mild and the condition discovered at the adult age. Conversely it can be severe, giving symptoms of hemolytic anemia in early infancy. Several cases of neo-natal hemolytic syndrome have been described. Usually there are no cytological abnormalities of the red cells, except a high degree of reticulocytosis. However, cases with distortion of the erythrocyte shape have been reported [43]. Splenectomy seems to bring a definite improvement in these severe forms with an important shortage of red cell life span. It is noteworthy that after splenectomy there is an increase in reticulocyte counts, suggesting a sieving effect of spleen on these cells [31]. In most cases there is an increase of *in vitro* auto-hemolysis not corrected by adenosine triphosphate (ATP). Thus pyruvate kinase deficiency belongs to the type II group of chronic hemolytic anemias as defined by SELWYN and DACIE on the basis of auto-hemolysis [57]. Actually this test, which had at one time an operational value for classification of CNHD, now appears to be almost obsolete. Its significance has been questioned many times. The puzzling correcting effect of ATP, a compound which cannot cross the erythrocyte membrane, has been recently reinvestigated by DACIE's group [21]. It was shown that ATP is only active through an acidification of the medium, and that the correction of increased auto-hemolysis is no longer observed when neutralized ATP is used.

The diagnosis of pyruvate kinase deficiency is ascertained by the enzyme assay in hemolysed red cells. This can be done by spectrophotometric measurement of NADH disappearance in the lactic dehydrogenase (LDH) coupled reaction [60]. Special care must be taken to avoid any contamination of red cells by leukocytes, since the pyruvate kinase activity of these cells is 300 times higher than that of normal red cells, and is not affected in pyruvate kinase deficiency [58]. The pyruvate kinase level in deficient red cells from homozygous subjects is usually diminished to 5 to 20% of the normal level [59]. In the heterozygotes, the residual activity is around 50%. These subjects do not display any clinical or hematological abnormality. It is noteworthy that there is no apparent correlation between the enzyme level and the degree of hemolysis in the patients. Two screening tests have been described. The first [13] is based upon the increase of pH which occurs during the pyruvate kinase reaction. A coloured pH indicator is used. The second [4] is much more specific because the consumption of NADH by the endogenous LDH coupled reaction is followed by examination of dry spots under UV light. The defluorescence rate parallels the pyruvate kinase activity. This test permits diagnosis of both homozygous and heterozygous states [50].

Metabolic abnormalities of pyruvate kinase deficient red cells have been extensively studied [14, 16, 20, 31, 59, 73]. Conflicting results have been reported as far as the ATP level is concerned [31, 59, 64], although a lowered ATP level would be expected since pyruvate kinase catalyses one of the two ATP producing steps in the red cell. As emphasized by KEITT [31] the high proportion of reticulocytes probably accounts for an extra-source of ATP through mitochondrial oxidative phosphorylation. The same reason would explain why glucose is, in some cases, metabolized at an almost normal rate. But when the reticulocyte content is taken into consideration, it becomes evident that glucose utilization is impaired in pyruvate kinase deficient red cells [59].

The accumulation of 2,3-DPG has been called the 'hallmark' of pyruvate kinase deficiency [31], since it is constantly found. It has been suggested that this phenomenon is due to an up-stream accumulation of glycolytic intermediates particularly 3-PG which would inhibit the first ATP producing reaction: phosphoglycerate kinase (PGK), forcing 1,3-PG to enter the wasteful Rapoport-Luebering cycle [14, 16, 76]. However, this coherent hypothesis has been questioned by ZUELZER *et al.* [75]. There is surprisingly good evidence that the pyruvate kinase reaction is by no means a rate limiting step in glucose metabolism in normal red cells, where the enzyme operates far below saturation with respect to phosphoenolpyruvate (PEP) [49]. Actually, despite a considerable amount of investigations, the metabolic disturbances caused by pyruvate kinase deficiency remain unclear, and the direct cause of hemolysis remains unknown as well.

In terms of molecular genetics, the disease is not better understood. The majority of reported kindreds offer a 'regular' pattern with low enzyme activity and full clinical expression in homozygotes, and intermediate enzyme activity without clinical symptoms in heterozygotes. This suggests a recessive, autosomal mode of inheritance. On the other hand, there are increasing reports of affected individuals with a pyruvate kinase level in the intermediate range [32, 44, 59], and considerable variation of pyruvate kinase values, as stressed by TANAKA [59], is now apparent *between* families as well as *within* families [42, 75]. This strongly suggests that a genetic polymorphism exists. Genetic analysis of some kindreds has shown that the affected subject is apparently a heterozygote [9, 50]. Other clues supporting a molecular polymorphism of the condition are those results concerning the search for kinetic abnormalities of the enzyme. Whereas some authors have found no abnormality of the Michaelis constant (K_M) for PEP [15, 49, 74], some others have reported high K_M values for this substrate [44, 50, 72]. This fact might lead one to overlook the enzyme deficiency in the regular optical assay system where there is a considerable excess of PEP. It has been suggested that all assays be run systematically at two concentrations of substrate: high and low, the latter being still saturating for the normal enzyme [44]. Recently BOIVIN *et al.* [7, 8], investigating seven homozygotes for pyruvate kinase deficiency, found that the K_M for PEP was normal in three cases, increased in two cases, and decreased in two cases. These findings support the idea that pyruvate kinase deficiency is, like G6PD deficiency, a heterogeneous condition from a molecular point of view. However, this has been questioned by CARTIER *et al.* [16] who found the red cell pyruvate kinase to exist in two interconvertible forms with different K_M for PEP, as already reported in adipose tissue [46]: an A form with allosteric kinetics and high K_M and a B form with regular Michaëlian kinetics and low K_M. Pyruvate kinase A is activated by fructose diphosphate (FDP) while its K_M for PEP becomes equal to that of pyruvate kinase B. If this is true, it casts a serious doubt upon the reported abnormalities of kinetics, since minute amounts of FDP may bring drastic changes of K_M for PEP values. Actually, there is a considerable discrepancy between normal data reported, probably owing to environmental differences. Extreme values such as 10^{-3} M [7, 33, 49] and 10^{-5}M [16, 24] have been reported.

The problem of the electrophoretic pattern of pyruvate kinase in deficient red cells is still not clarified. Its study is hampered by technical difficulties, since a defluorescence method has to be used [71]. Two bands have been found in normal red cells [6, 63]. In pyruvate kinase deficiency both bands were reported as missing in some cases [63], whereas in other cases only one of them was weak and displaced, the other remaining normal [6]. These discrepancies make it presently impossible to decide whether or not both bands are a single gene product.

As a conclusion, we can say that there are now many facts which indicate that polymorphism exists at the genetic, i.e. molecular level. To reach certitude, we must await more information about the properties of pyruvate kinase, an enzyme which has never been completely purified as yet.

B. Triose Phosphate Isomerase

Triose phosphate isomerase (TPI) deficiency has been found to be a cause of congenital hemolytic anemia by SCHNEIDER *et al.* [51]. Since this first report, the same group of investigators has collected three kindreds in which a high degree of consanguinity was found [52, 66]. Three homozygotes were studied, and four additional homozygotes were highly suspected on the basis of the clinical pattern which is strinkingly similar in all affected subjects. The disorder is characterized by a severe congenital hemolytic anemia recognized at birth or few weeks later. Reticulocyte counts are around 15%, without gross morphological abnormalities of red cells. It is of considerable interest that, in all cases, symptoms of neurologic involvement appeared after six months of age, with progressive spasticity ending in flaccid paralysis. The neurologic syndrom was not clearly defined, but was remarkable for its progressiveness and age of appearance. Also noteworthy is an increased susceptibility to infection. All but two subjects died, either very early from acute hemolysis, or later from severe infection, or in two cases suddenly without any immediate overt cause.

A great number of clinically unaffected heterozygotes have been identified in the kindreds permitting the establishment of a typical autosomal recessive mode of inheritance. The trait was found first in individuals of French and African ancestry living in Louisiana and belonging to highly inbred families. Later it was found in one subject from English ancestry, and in one Negro [52].

The basic enzyme defect was identified in the three subjects available as a considerable loss of TPI activity in the red cells. The residual activity was around 7% in two subjects and somewhat higher, around 35% in the third one [52]. The leukocytes were also deficient, although to a smaller extent [52]. Moreover, in one well-documented individual, the enzyme activity was found to be diminished not only in blood cells, but also wherever it had been assayed: muscle, cerebrospinal fluid and serum [52]. These findings suggest that the enzyme deficiency is widespread in the body, an assumption which is in accordance with the variety of extrahematological symptoms.

The diagnosis of TPI deficiency can be easily achieved by an optical method, in which the generated dihydroxy-acetone phosphate (DHAP) is reduced by added α-glycerophosphate-dehydrogenase in the presence of NADH [51]. It should be noted that in the original report the substrate (phosphoglyceraldehyde) concentration used was sevenfold below the saturating level [51]. With this method the homozygotes exhibit an important decrease of activity. The heterozygotes are in the intermediate range, and usually the enzyme assay provides a good means for the detection of carriers, who are clinically unaffected.

It is striking that the enzyme level is very low in homozygotes although the red cell population is quite young. This is in contrast with many other enzyme deficiencies such as pyruvate kinase or hexokinase in which the deficiency is offset by the young age of the average red cell population. This is possibly due to the fact that TPI activity does not change very much during erythrocyte aging [41].

We have devised a fast screening method for detection of TPI deficiency [26]. It is based on the same principle as the original UV spot-test described by BEUTLER [4]. It permits to discriminate easily not only homozygotes, but also unaffected heterozygous carriers [26].

Metabolic abnormalities have been well-documented in two cases [53]. Surprisingly enough, glucose consumption and lactate production were found to be augmented even if compared to a high reticulocyte control. The ATP and 2,3-DPG content were normal. The most prominent abnormality was a tenfold increase of DHAP as compared to the high reticulocyte control. There is good evidence that the HMP shunt works at its maximal capacity. How is the red cell capable of maintaining a normal ATP content despite a major defect such as TPI deficiency? This remains unclear, although it has been said that the HMP shunt might provide a salvage pathway [53]. Moreover, whereas the basic defect is undoubtedly the TPI deficiency, nothing is known regarding the immediate cause of hemolysis. A deleterious effect of the accumulated DHAP has been hypothesized [53], since the red cell, which has no α-glycerophosphate dehydrogenase, cannot metabolize this substrate. Conversely, it has been speculated that in other tissues, particularly in the nervous system, the TPI deficiency might produce a trouble in glycerol synthesis through a shortage in DHAP [53].

The mechanism of the decrease of enzyme activity in TPI deficiency remains to be defined. No inhibitor is responsible [51, 66]. No abnormal kinetics concerning glyceraldehyde-3-phosphate as a substrate were noted [51]. In fact, data are too scarce to allow a definite conclusion regarding the hypothesis of a structurally abnormal allel. We have attempted to solve this problem by means of an electrophoretic separation followed by specific localization on starch gel [27]. Three major defluorescing bands were seen in human red blood cells as well as in different tissues from rabbit. In the red cells of one subject with TPI deficiency the fastest band was absent, the intermediate band barely visible and the slowest band diminished but to a smaller extent. Different speculations were made about the interpretation of these results [27].

C. Glyceraldehyde-3-phosphate Dehydrogenase

Glyceraldehyde-3-phosphate dehydrogenase (GAPD) deficiency has been assumed to be the cause of chronic hemolytic disease in two individuals, father and son, of English ancestry [22]. The enzyme content of red cells was 20 to 30% of normal. Biochemical abnormalities involving fructose-1,6-diphosphate (FDP), ATP and 2,3-DPG were found. The alterations were of the same type as those produced in normal cells by iodoacetate, a specific inhibitor of GAPD. No further information has appeared about this family since the first preliminary report in 1966. Therefore, this unique type of deficiency still awaits confirmation.

D. Hexokinase

Red cell hexokinase (HK) deficiency has been observed in only one kindred [67, 68]. In the single homozygous individual described, it resulted in a chronic hemolytic anemia which was already manifest at birth, necessitating exchange transfusion. Despite a splenectomy, the hemolytic process remained severe, with variable jaundice and reticulocytosis. The half-life of cells tagged with chromium was reduced to six days. Enzyme assays demonstrated a relative decrease of red cell hexokinase activity. Actually, this apparently borderline abnormality is of great significance if the young age of the red cell population is taken into consideration, since there is a several fold decrease of hexokinase between the reticulocyte and the aged cell [11, 41]. VALENTINE *et al.* [67, 68] have shown that, according to the reticulocyte count and the computed mean age of the red cells, the hexokinase should have been several times higher in their patient. This is also exemplified when the ratios hexokinase versus pyruvate kinase and hexokinase versus G-6-PD are calculated. In contrast with high reticulocyte control red cells, these deficient cells display a sharp decrease in their hexokinase versus pyruvate kinase and hexokinase versus G-6-PD ratios [68]. Glucose and fructose were consumed at less than normal rates. Mannose, in contrast, was actively metabolized. This phenomenon was attributed to a separate, unaffected kinase [68]. But this interpretation was questioned recently when it appeared that the phosphomannose isomerase activity is very high and no longer rate-limiting in young erythrocytes [5]. In hexokinase deficient red cells a higher proportion of glucose was found to be metabolized via the hexose monophosphate shunt. It is of great interest that the enzyme deficiency was not found in leukocytes and platelets [67]. This is not surprising, since we now know that hexokinase can be resolved into several isoenzymes which may be identified by electrophoresis [30]. There is a current controversy about the normal pattern and signification of red cell hexokinase isoenzymes [23, 28, 29, 40, 56]. A normal pattern was found in the hexokinase deficient red cells where the bands were only reduced in intensity [68]. Enzyme kinetics with glucose and ATP were found to be normal. In the same family, the parents and three siblings were assumed to be heterozygotes. Among them the red cell hexokinase level was strikingly scattered: almost normal in the mother, and even lower than in the proposita in one brother. Since none of these subjects had hematological abnormalities, it is reasonable to speculate that their middle age red cells could cope with the decreased hexokinase activity whereas the patient's very young red cells suffered more from the deficiency [68]. It is also possible that the *in vivo* stability of the enzyme is highly impaired in deficient red cells resulting in a premature hexokinase decay and cellular death below a certain level of enzyme activity[1].

E. Phospho-hexose Isomerase

Phospho-hexose isomerase (PHI) deficiency is a recently discovered condition giving rise to a severe hemolytic anemia. It was first described by the same group of authors in two unrelated families [1, 45]. A third family has been very recently discovered in France [18]. Despite marked reticulocytosis, the PHI activity ranged from 14 to 40% in the five homozygous patients hitherto reported. Red cells from parents and some siblings had an intermediate activity, without clinical symptoms, suggesting a regular autosomal recessive mode of inheritance. The defect was found to be even more pronounced in leukocytes [1, 45]. However, in the third family the leukocyte activity was only reduced to 50% [18]. In contrast with TPI deficiency, no extra-hematologic involvement was detectable in the patients. Again, in this disease there is no appreciable decrease of glycolysis in red cells. Glucose consumption was even found to be increased [18], as was the ATP level [1]. Glucose-6-phosphate (G6P) does not accumulate in contrast with what one might have anticipated. The only metabolic defect reflecting the enzyme deficiency is the loss of recycling capacity for substrates coming from the HMP shunt [1]. But the metabolic reason why the life span is so drastically shortened in PHI deficient red cells still remains obscure [18]. Electrophoresis of PHI in the three families has provided interesting results. An abnormal pattern was found in family I [19]. In family II the electrophoretic mobility was also modified but in a different way [45], while in family III no abnormality was detected [18]. This strongly suggests that at least in family I and II, the deficiency is due to a structurally abnormal protein. This has been well-documented in family I where the patient appears to be heterozygous for two different rare alleles at the PHI locus [45]. It is likely that genetic

[1] In 1965, LÖHR *et al.* [37] described a diminished hexokinase level in red cells, leukocytes and platelets in three children with Fanconi's syndrome. It was associated with chromosomal aberrations and is certainly a separate morbid entity, distinct from the one depicted above.

polymorphism exists in PHI deficiency since the electrophoretic data collected from the three affected families are different. The geographical origin of these kindreds is also different suggesting that the condition might be relatively widespread.

F. Phosphoglycerate Kinase

Phosphoglycerate Kinase (PGK) Deficiency. In 1968 two groups have almost simultaneously reported on red cell phosphoglycerate kinase deficiency as a cause for CNHD [34, 69]. The first report [34] concerned a 63-year-old female patient, with long lasting hemolytic disease, and a ^{51}Cr red cell half survival of 12 days. All enzymes tested were high, except PGK which was slightly diminished. Young cells were shown to contain a normal amount of PGK activity, whereas relatively old cells contained only 25% of residual activity. The leukocytes were normal. No genetic study could be made in the absence of other living members of the family.

In contrast, the second report [69] concerns a very large kindred in which the deficiency could be well-documented. In a Chinese family, two male children were observed with a severe hemolytic disease associated to neurological abnormalities. The PGK activity was reduced to almost nothing in the red cells. In the leukocytes the residual activity was around 5%. In the same family five males had previously died in infancy with anemia of unknown origin and neurological disorders.

Of great interest is the fact that the genealogy of this kindred, which comprises three generations available for study, gave strong evidence for X-chromosome linkage. This assumption was supported by a special study of the heterozygous females. Some, with mild PGK deficiency, exhibited a slight degree of hemolysis. Their young red cells had a lower PGK activity than old cells, a finding which suggests that there are two red cell populations, one with the trait and a short life span, the other without the trait and a normal survival [69, 70]. These results are quite consistent with the Lyon-Beutler theory of mosaicism [3, 39]. Thus, the affected males are probably hemizygotes. Their red cells can survive shortly because the Embden-Meyerhof pathway still operates in the very young cells. ATP is decreased as compared to the ATP level in young normal red cells, and 2,3-DPG is markedly elevated like in pyruvate kinase deficiency. It is very likely that the Rapoport-Luebering by-pass is the main metabolic route in these cells.

It is obvious that KRAUS' and VALENTINE's observations [34, 69] are different, and even opposite. Genetic polymorphism may be the cause of these discrepancies. At the present time there are good indications that PGK might be another X-chromosome marker.

G. Phosphofructokinase

Phosphofructokinase (PFK) deficiency is the cause of a new type of glycogenosis which was identified recently in three adult siblings [61]. It is now designated as type VII glycogen-storage disease [12]. Aside from the muscular disease there are signs of hematologic involvement: moderate wellcompensated hemolytic disease without spherocytosis, slightly increased reticulocyte counts, and shortened life span of ^{51}Cr labelled erythrocytes which is around 15 days [62]. This condition is quite rare since only two affected kindreds have been published until now [35, 61]. However, it offers a nice example of how a clinical disorder can shed some light on basic problems of molecular genetics. It has been shown that, although the enzyme deficiency is virtually complete in skeletal muscle, there is only a 50% decrease in erythrocytes [61]. Immunological studies and chromatographic separation have demonstrated that 50% of the normal erythrocyte PFK activity is due to a 'muscle type' PFK protein [62]. It is, therefore, evident that in erythrocytes there are at least two PFK isoenzymes which are not allelic. If the gene governing the 'muscle type' is abnormal, the other isoenzyme is not affected in erythrocytes. On the other hand, a reduction of 50% of the PFK activity appears to be deleterious for the erythrocyte. This is in accordance with its role as a key enzyme, a bottle-neck in the red cell glycolytic pathway [17].

II. Enzymes of the Rapoport-Luebering Cycle

A. 2,3-Diphosphoglycerate Mutase

A deficiency of 2,3-diphosphoglycerate mutase (DPGM) was suspected by PRANKERD in two subjects with mild CNHD [10, 47]. The enzyme deficiency was assumed on an indirect basis, consisting of the demonstration of low levels of 2,3-diphosphoglyceric acid (2,3-PGA) in the red cells. These were also found to be unable to synthesize this compound after exhaustion of 2,3-PGA or incubation with fluoride which, in normal cells, induces an accumulation of 2,3-PGA. As the disease was found in a father and his son the condition was assumed to be transmitted as a dominant trait.

LÖHR and WALLER [36, 72] made similar observations in several patients. They also brought indirect evidence for the DPGM deficiency. In the red cells of their patients, lactate production was found impaired, ATP was decreased, and methemoglobin detectable. The authors claimed that in the absence of 2,3-DPG which is a coenzyme of the monophosphoglycerate mutase reaction, this step, located on the main glycolytic pathway, is blocked.

Indirect evidence such as the inability of the red cell to synthesize 2,3-PGA in various conditions is not specific for DPGM deficiency, since TPI deficiency can realize the same biochemical pattern. Therefore, the only specific way of diagnosis would be a direct assay of DPGM. Unfortunately, at the present time there is no direct assay available. Even the assay devised by SCHRÖTER *et al.* [55] is somewhat indirect. It has been used in one family in which a child had died from severe hemolytic anemia in early infancy [54]. No enzyme measurement was performed in the propositus. In the parents and two relatives the DPGM was found to be diminished by 50% in the red cells. It was assumed that the patient was homozygous for DPGM deficiency and that the condition is inherited as a recessive trait.

It thus appears that DPGM deficiency is not clearly defined as yet, since no direct measurement of the enzyme activity could ever be made in a patient. The discovery by BENESCH [2] that 2,3-PGA exerts a regulatory effect on hemoglobin oxygen affinity might be an interesting line of development for the study of this enzyme deficiency.

B. 2,3-Diphosphoglycerate Phosphatase (DPGase)

JACOBASH *et al.* [25] have reported on two subjects with mild hemolytic syndrome associated to neuromuscular disorders and hyperlipemia. The red cells contained high ATP level and no DPGase activity. This deficiency, which was assumed to be the basic cause of the increase of ATP, does not provide, however, any explanation for the accumulation of *all* adenylic nucleotides. More studies are needed to clarify this very peculiar syndrom, which is presumably different from the other conditions in which a high ATP level has been found in the erythrocyte [38].

Conclusion

We have seen that there are now a good number of enzyme deficiencies giving rise to a hemolytic disease. This does not preclude the possible discovery of additional red cell enzyme abnormalities in the future.

It must be emphasized that the decrease of a given enzyme activity does not necessarily mean that this represents the basic cause of the disease. Nor does it give an indication as to the mechanism of red cell destruction. Another problem is the molecular significance of the enzyme deficiency. This question is not particular to the red cell and we will not attempt to give all possible answers. Let us say that, because of its availability, the red cell represents a unique model for the study of some molecular diseases. Despite its inability to divide, this cell appears to be as useful to the medical scientist as *Escherichia coli* is to the molecular geneticist.

References

1. BAUGHAN, M. A.; VALENTINE, W. N.; PAGLIA, D. E.; WAYS, P. O.; SIMONS, E. R. and DEMARSH, Q. B.: Hereditary hemolytic anemia associated with glucose phosphate isomerase (GPI) deficiency. A new enzyme defect of human erythrocytes. Blood *32:* 236 (1968).
2. BENESCH, R. and BENESCH, R. E.: Intracellular organic phosphates as regulators of oxygen release by haemoglobin. Nature, Lond. *221:* 618 (1969).
3. BEUTLER, E.; YEH, M. and FAIRBANKS, V. E.: The normal human female as a mosaic of X-chromosome activity: studies using the gene for G6PD deficiency as a marker. Proc. nat. Acad. Sci. *48:* 9–16 (1962).
4. BEUTLER, E.: A series of new screening procedures for pyruvate kinase deficiency, glucose-6-phosphate dehydrogenase deficiency and glutathione reductase deficiency. Blood *28:* 553 (1966).
5. BEUTLER, E. and TEEPLE, L.: Mannose metabolism in the human erythrocyte. J. clin. Invest. *48:* 461 (1969).
6. BLUME, K. G.; LOHR, G. W.; RUDIGER, G. W. and SCHALHORN, A.: Pyruvate kinase in human erythrocytes. Lancet *i:* 529 (1968).
7. BOIVIN, P. et GALLAND, C.: Recherche d'une anomalie moléculaire lors des déficits en pyruvate kinase érythrocytaire. Nouv. Rev. franç. Hémat. *8:* 201 (1968).
8. BOIVIN, P. and GALLAND, C.: Variants of erythrocyte pyruvate kinase. Lancet *ii:* 356 (1968).
9. BOSSU, M.; DACHA, M. and FORNAINI, G.: Neonatal hemolysis due to a transient severity of inherited pyruvate kinase deficiency. Acta haemat. *40:* 166 (1968).
10. BOWDLER, A. J. and PRANKERD, T. A. J.: Studies in congenital non-spherocytic haemolytic anemias with special enzyme defects. Acta haemat. *31:* 65 (1964).
11. BROK, F.; RAMOT, B.; ZWANG, E. and DANON, D.: Enzyme activities in human red blood cells of different age groups. Israel J. med. Sci. *2:* 291 (1966).
12. BROWN, B. I. and BROWN, D. H.: In F. DICKENS, P. J. RANDLE and W. J. WHELAN Carbohydrate metabolism and its disorders, vol. 2, p. 144 (Academic Press, London/New York 1968).
13. BRUNETTI, P. and NENCI, G.: A screening method for the detection of erythrocyte pyruvate kinase deficiency. Enzym. biol. clin. *4:* 51 (1964).

14. BUSCH, D.: Probleme des Erythrozytenstoffwechsels bei Anämien mit Pyruvatkinasemangel. Folia haemat. *83:* 395 (1965).

15. CAMPOS, J. O.; KOLER, R. D. and BIGLEY, R. H.: Kinetic differences between human red cell and leukocyte pyruvate kinase. Nature, Lond. *208:* 194 (1965).

16. CARTIER, P.; NAJMAN, A.; LEROUX, J. P. et TEMKINE, H.: Les anomalies de la glycolyse au cours de l'anémie hémolytique par déficit du globule rouge en pyruvate kinase. Clin. chim. Acta *22:* 165 (1968).

17. CARTIER, P.: La glycolyse du globule rouge normal et pathologique. Exp. Ann. Biochim. Méd., vol. 29, p. 25 (Masson, Paris 1969).

18. CARTIER, P.; TEMKINE, H. et GRISCELLI, C.: Etude biochimique d'une anémie hémolytique avec déficit familial en phosphohexo-isomérase. (To be published.)

19. DETTER, J. C.; WAYS, P. O.; GIBLETT, E. R.; BAUGHAN, M. A.; HOPKINS, D. A.; POVEY, S. and HARRIS, H.: Inherited variations in human phosphohexose isomerase Ann. hum. Genet. *31:* 329 (1968).

20. GRIMES, A. J.; MEISLER, A. and DACIE, J. V.: Hereditary non-spherocytic haemolytic anemia. A study of red cell carbohydrate metabolism in twelve cases of pyruvate kinase deficiency. Brit. J. Haemat. *10:* 403 (1964).

21. GRIMES, A. J.; LEETS, I. and DACIE, J. V.: The autohaemolysis test: appraisal of the method for the diagnosis of pyruvate kinase deficiency and the effects of pH and additives. Brit. J. Haemat. *14:* 309 (1968).

22. HARKNESS, D. R.: A new erythrocytic enzyme defect with hemolytic anemia: glyceraldehyde-3-phosphate-dehydrogenase deficiency. J. Lab. clin. Med. *68:* 879 (1966).

23. HOLMES, E.; MALONE, J.; WINEGRAD, A. and OSKI, F. A.: Hexokinase isoenzymes in human erythrocytes: association of type II with fetal hemoglobin. Science *156:* 646 (1967).

24. IBSEN, K. H.; SCHILLER, K. W. and VENN-WATSON, E. A.: Stabilization, partial purification and effects of activating cations ADP and phosphoenolpyruvate on the reaction rates of an erythrocyte pyruvate kinase. Arch. Biochem. Biophys. *128:* 583 (1968).

25. JACOBASCH, G.; SYLLM-RAPOPORT, I.; ROIGAS, H. und RAPOPORT, S.: 2,3-PGase-Mangel als mögliche Ursache erhöhten ATP-Gehaltes. Clin. chim. Acta *10:* 477 (1964).

26. KAPLAN, J. C.; SHORE, H. and BEUTLER, E.: The rapid detection of triose phosphate isomerase deficiency. Amer. J. clin. Path. *50:* 655 (1968).

27. KAPLAN, J. C.; TEEPLE, L.; SHORE, H. and BEUTLER, E.: Electrophoretic abnormality in triose phosphate isomerase deficiency. Biochem. biophys. Res. Comm. *31:* 768 (1968).

28. KAPLAN, J. C. and BEUTLER, E.: Hexokinase isoenzymes in human erythrocytes. Science *159:* 215 (1968).

29. KAPLAN, J. C. and BEUTLER, E.: Hexokinase isoenzymes. New Engl. J. Med. *280:*1,129 (1969).

30. KATZEN, H. M.: The multiple forms of mammalian hexokinase and their significance to the action of insulin; in G. WEBER, Advances in enzyme regulation, vol. 5, p. 335 (Pergamon Press, New York 1967).

31. KEITT, A. S.: Pyruvate kinase deficiency and related disorders of red cell glycolysis. Amer. J. Med. *41:* 762 (1966).

32. KOLER, R. D.; BIGLEY, R. H.; JONES, R. T.; RIGAS, D. A.; VANBELLINGHEN, P. and THOMPSON, P.: Pyruvate kinase: molecular differences between human red cell and leucocyte enzyme. Cold Spr. Harb. Symp. quant. Biol. *29:* 213 (1964).
33. KOLER, R. D.; BIGLEY, R. H. and STENZEL, P.: Biochemical properties of human erythrocyte and leucocyte pyruvate kinase; in E. BEUTLER, Hereditary disorders of erythrocyte metabolism, City of Hope Symposium Series, vol. 1, p. 249 (Grune and Stratton, New York 1968).
34. KRAUS, A. P.; LANGSTON, M. F. Jr. and LYNCH, B. L.: Red cell phosphoglycerate kinase deficiency. A new cause of non spherocytic hemolytic anemia. Biochem. biophys. Res. Comm. *30:* 173 (1968).
35. LAYZER, R. B.; ROWLAND, L. P. and RANNEY, H. M.: Muscle phosphofructokinase deficiency. Arch. Neurol. *17:* 512 (1967).
36. LÖHR, G. W. und WALLER, H. D.: Zur Biochemie einiger angeborener hämolytischer Anämien. Folia haemat. *8:* 377 (1963).
37. LÖHR, G. W.; WALLER, H. D.; ANSCHUTZ, F. und KNOPP, A.: Biochemische Defekte in den Blutzellen bei familiärer Pan-Myelopathie (Type Fanconi). Humangenetik *I:* 383 (1965).
38. LOSS, J. A.; PRINS, H. K. and ZURCHER, C.: Elevated ATP levels in human erythrocytes; in E. BEUTLER, Hereditary disorders of erythrocyte metabolism, p. 280. City of Hope Symposium Series, vol. 1 (Grune and Stratton, New York 1968).
39. LYON, M. F.: Gene action in the X-chromosome of the mouse (*Mus musculus*). Nature, Lond. *190:* 372 (1961).
40. MALONE, J. I.; WINEGRAD, A. I.; OSKI, F. A. and HOLMES, E. W.: Erythrocyte hexokinase isoenzyme patterns in hereditary hemoglobinopathies. New Engl. J. Med. *279:* 1,071 (1968).
41. MARCHAND, J. C. et GARREAU, H.: L'activité des enzymes glycolytiques du réticulocyte et du globule rouge en fonction de l'âge. C. R. Soc. Biol. *162:* 1,302 (1968).
42. OSKI, F. A. and DIAMOND, I. K.: Erythrocyte pyruvate kinase deficiency resulting in congenital non-spherocytic hemolytic anemia. New Engl. J. Med. *269:* 763 (1963).
43. OSKI, F. A.; NATHAN, D. G.; SIDEL, V. W. and DIAMOND, L. K.: Extreme hemolysis and red cell distortion in erythrocyte pyruvate kinase deficiency. I. Morphology, erythrokinetics and family enzymes studies. New Engl. J. Med. *270:* 1,023 (1964).
44. PAGLIA, D. E.; VALENTINE, W. N.; BAUGHAN, M. A.; MILLER, D. R.; REED, C. F. and MCINTYRE, O. R.: An inherited molecular lesion of erythrocyte pyruvate kinase. J. clin. Invest. *47:* 1929 (1968).
45. PAGLIA, D. E.; HOLLAND, P.; BAUGHAN, M. A. and VALENTINE, W. N.: Occurrence of defective hexosephosphate isomerization in human erythrocytes and leucocytes. New Engl. J. Med. *280:* 66 (1969).
46. POGSON, C. I.: Adipose tissue pyruvate kinase. Properties and interconversion of two active forms. Biochem. J. *110:* 67 (1968).
47. PRANKERD, T. A. J.: In H. SCHUBOTHE Hämolyse und hämolytische Erkrankungen. VII. Freiburger Symp., 1959, p. 136 (Springer, Berlin 1961).
48. PRANKERD, T. A. J.: Inherited enzyme defects in congenital hemolytic anemias. Proc. 9th Congr. Europ. Soc. Haemat., Lisbon 1963, p. 735 (Karger, Basel/New York 1963).
49. ROSE, I. A. and WARMS, J. V. B.: Control of glycolysis in the human red blood cell. J. biol. Chem. *241:* 4,848 (1966).

50. SACHS, J. R.; WICKER, D. J.; GILCHER, R. O.; CONRAD, M. E. and COHEN, R. J.: Familial hemolytic anemia resulting from an abnormal red blood cell pyruvate kinase. J. Lab. clin. Med. *72:* 359 (1968).

51. SCHNEIDER, A. S.; VALENTINE, W. N.; HATTORI, M. and HEINS, H. L., Jr.: Hereditary hemolytic anemia with triose phosphate isomerase deficiency. New Engl. J. Med. *272:* 229 (1965).

52. SCHNEIDER, A. S.; VALENTINE, W. N.; BAUGHAN, M. A.; PAGLIA, D. E.; SHORE, N. A. and HEINS, H. L., Jr.: Triose phosphate isomerase deficiency. A. A multi-system inherited enzyme disorder. Clinical and genetic aspects; in E. BEUTLER, Hereditary disorders of erythrocyte metabolism, p. 265. City of Hope Symposium Series, vol. 1 (Grune and Stratton, New York 1968).

53. SCHNEIDER, A. S.; DUNN, I.; IBSEN, K. H. and WEINSTEIN, I. M.: Triose phosphate isomerase deficiency. B. Inherited triose phosphate isomerase deficiency. Erythrocyte carbohydrate metabolism and preliminary studies of the erythrocyte enzyme; in E. BEUTLER, Hereditary disorders of erythrocyte metabolism, p. 273. City of Hope Symposium Series, vol. 1 (Grune and Stratton, New York 1968).

54. SCHRÖTER, W.: Kongenitale nichtsphärocytäre hämolytische Anämie bei 2,3-Diphosphoglyceratmutase-Mangel der Erythrocyten im frühen Säuglingsalter. Klin. Wschr. *43:* 1,147 (1965).

55. SCHRÖTER, W. und HEYDEN, H.: Kinetic des 2,3-Diphosphoglyceratumsatzes in menschlichen Erythrocyten. Biochem. Z. *341:* 387 (1965).

56. SCHRÖTER, W. and TILLMANN, W.: Hexokinase isoenzymes in human erythrocytes of adults and newborns. Biochem. biophys. Res. Comm. *31:* 92 (1968).

57. SELWYN, J. G. and DACIE, J. V.: Autohemolysis and other changes resulting from the incubation *in vitro* of red cells from patients with congenital hemolytic anemia. Blood *9:* 414 (1954).

58. TANAKA, K. R.; VALENTINE, W. N. and MIWA, S.: Pyruvate kinase (PK) deficiency in hereditary non-spherocytic hemolytic anemia. Blood *19:* 267 (1962).

59. TANAKA, K. R. and VALENTINE, W. N.: Pyruvate kinase: in E. BEUTLER, Hereditary disorders of erythrocyte metabolism, p. 229. City of Hope Symposium Series, vol. 1 (Grune and Stratton, New York 1968).

60. TANAKA, K. R.: Pyruvate kinase; in J. J. YUNIS, Biochemical methods in red cell genetics, p. 167 (Academic Press, New York 1969).

61. TARUI, S.; OKUNO, G.; IKURA, Y.; TANAKA, T.; SUDA, M. and NISHIKAWA, M.: Phosphofructokinase deficiency in skeletal muscle. A new type of glycogenosis. Biochem. biophys. Res. Comm. *19:* 517 (1965).

62. TARUI, S.; KONO, N.; NASU, T. and NISHIKAWA, M.: Enzymatic basis for the coexistence of myopathy and hemolytic disease in inherited muscle phosphofructokinase deficiency. Biochem. biophys. Res. Comm. *34:* 77 (1969).

63. TOWNES, P. L.: In E. BEUTLER, Hereditary disorders of erythrocyte metabolism, p. 259. City of Hope Symposium Series, vol. 1 (Grune and Stratton, New York 1968).

64. TWOMEY, J.; O'NEAL, F. B.; ALFREY, C. P. and MOSER, R. H.: ATP metabolism in pyruvate kinase deficient erythrocytes. Blood *30:* 576 (1967).

65. VALENTINE, W. N.; TANAKA, K. R. and MIWA, S.: A specific erythrocyte glycolytic enzyme defect (pyruvate kinase) in three subjects with congenital non-spherocytic hemolytic anemia. Trans. Ass. Amer. Physicians *74:* 100 (1961).

66. VALENTINE, W. N.; SCHNEIDER, A. S.; BAUGHAN, MA.; PAGLIA, D. E. and HEINS, H. L., Jr.: Hereditary hemolytic anemia with triose phosphate isomerase deficiency. Studies in kindreds with coexistent sickle cell trait and erythrocyte glucose-6-phosphate-dehydrogenase deficiency. Amer. J. Med. *41:* 27 (1966).
67. VALENTINE, W. N.; OSKI, F. A.; PAGLIA, D. E.; BAUGHAN, M. A.; SCHNEIDER, A. S. and NAIMAN, J. L.: Hereditary hemolytic anemia with hexokinase deficiency. New Engl. J. Med. *276:* 1 (1967).
68. VALENTINE, W. N.; OSKI, F. A.; PAGLIA, D. E.; BAUGHAN, M. A. SCHNEIDER, A. S. and NAIMAN, J. L.: Erythrocyte hexokinase and hereditary hemolytic anemia; in E. BEUTLER, Hereditary disorders of erythrocyte metabolism, p. 288. City of Hope Symposium Series, vol. 1 (Grune and Stratton, New York 1968).
69. VALENTINE, W. N.; HSIEH, H. S.; PAGLIA, D. E.; ANDERSON, H. M.; BAUGHAN, M. A.; JAFFE, E. R. and GARSON, O. M.: Hereditary hemolytic anemia; association with phosphoglycerate kinase deficiency in erythrocytes and leucocytes. Trans. Ass. Amer. Physicians *81:* 49 (1968).
70. VALENTINE, W. N.; HSIEH, H. S.; PAGLIA, D. E.; ANDERSON, H. M.; BAUGHAN, M. A.; JAFFE, E. R. and GARSON, O. M.: Hereditary hemolytic associated with phosphoglycerate kinase in erythrocytes and leucocytes. New Engl. J. Med. *280:* 528 (1969).
71. VON FELLENBERG, R.; RICHTERICH, R. und AEBI, H.: Electrophoretisch verschieden wandernde Pyruvat-Kinasen aus einigen Organen der Ratte. Enzymol. biol. clin. *3:* 240 (1963).
72. WALLER, H. D. and LÖHR, G. W.: Hereditary non-spherocytic enzymopenic anemias with pyruvate kinase and 2,3-diphosphoglycerate mutase deficiency. IXth Congr. Europ. Soc. Hemat., Lisbon 1963.
73. WALLER, H. G. and LÖHR, G. W.: Hereditary non-spherocytic enzymopenic hemolytic anemia with pyruvate kinase deficiency. Proc. IXth Congr. Int. Soc. Hemat., Mexico City 1964, p. 257.
74. WIESMANN, N. U. and TÖNZ, O.: Investigations of the kinetics of red cell pyruvate kinase in normal individuals and in a patient with pyruvate kinase deficiency. Nature, Lond. *209:* 612 (1966).
75. ZUELZER, W. Z.; ROBINSON, A. R. and HSU, T. H.: Erythrocyte pyruvate kinase deficiency in non-spherocytic hemolytic anemia: a system of multiple genetic markers? Blood *32:* 33 (1968).
76. ZURCHER, D.; LOOS, J. A. and PRINS, H. K.: Hereditary high ATP content of human erythrocytes. Folia haemat. *83:* 366 (1965).

Authors' addresses: Prof. J. C. KAPLAN, Institut de Pathologie Moléculaire, 24, rue du Fg.-St-Jacques, *75 Paris 14*e, and Dr. CHRISTIANE KISSIN, Hôpital Debrousse, 23, rue Sœur-Bouvier, *69 Lyon 5*e (France)

7th int. Congr. clin. Chem., Geneva/Evian 1969; vol. 2: Clinical Enzymology, pp. 19–22
(Karger, Basel/München/Paris/New York 1970)

Methaemoglobinaemia

T. A. J. PRANKERD
University College Hospital Medical School, London

Haemoglobin is an unstable protein and readily undergoes spontaneous oxidation. In normal red cells, however, the concentration of methaemoglobin does not exceed 1% and this steady state can only be maintained by the presence of a continuous reductive mechanism in the cell. A number of ways in which methaemoglobin reduction can be brought about are listed in table I. The non-enzymatic processes are slow and probably do not contribute very much *in vivo*. Of the two enzymatic processes evidence suggests that reduction through an electron chain linked with NADH is much the more active. The presence of enzymes reducing methaemoglobin was suggested by the work of KIESE [6] and GIBSON [2], and attempts have been made to purify the enzymes responsible. SCOTT and MCGRAW [8] partially purified a flavin-containing protein which had a high affinity for NADH and only one tenth for NADPH. HUENNEKENS *et al.* [3] isolated three fractions with reductase activity, all of which were more active with NADH than with NADPH, and were thought to contain haem but no flavin compounds.

Table I. Mechanisms of methaemoglobin reduction

Non-enzymatic *slow*	Cysteine Ergothioneine	Reduced glutathione Ascorbic acid
Enzymatic *fast*	Methaemoglobin reductase NADH linked 10 NADPH linked 1	

NADH is formed in the red cell from NAD at the triose phosphate dehydrogenese reaction whilst NADPH is formed during the oxidation of glucose-6-phosphate in the pentose phosphate pathway. *In vitro*, both lactate and nucleosides will act as substrates for methaemoglobin reduction in addition to glucose, the former by supplying NADH during the conversion of lactate to pyruvate, and the latter by supplying substrate for the triose phosphate dehydrogenese after oxidation in the pentose shunt. That NADH is the main pathway for methaemoglobin reduction is evidenced by the absence of methaemoglobinaemia in patients with G-6-PD deficiency; it is possible, however, that the NADPH linkup may provide a supplementary route of reduction DE LOECKER and PRANKERD [1]. The well-known action of methylene blue in alleviating methaemoglobinaemia appears to be due to its bridging the electron gap between NADPH and methaemoglobin without the additional agency of an enzyme.

WEST [9] has recently achieved a 20,000 fold purification of methaemoglobin reductase and found a M.W. of about 30,000. This enzyme was ten times more reactive with NADH than NADPH and contained no flavin or haem group. Electrophoresis revealed a single protein band in the haemolysate but this dissociated into three bands after concentration or storage. Electrophoresis in the presence of NADH produced two bands which exhibited different K_M with NADH. Thus, there appeared to be two species of methaemoglobin reductase enzymes in human red cells both more reactive with NADH than NADPH.

Methaemoglobinaemia is not a common clinical manifestation of disease but its various causes are listed in table II. The mechanism through which various chemical agents produce methaemoglobulin is obscure, but those which act only *in vivo* probably do so as a result of the production of an intermediary product in the body. To protect haemoglobin against oxidation by these agents a number of reactions are possible in the red cell. Catalase and glutathione peroxidase break down hydrogen peroxide which is probably liberated by a number of agents interacting with haemoglobin, while the high concentration of reduced glutathione may provide a first line defence against oxidation through its readily available SH-groups.

In spite of these protective mechanisms and the presence of active methaemoglobin reduction in the cell, certain haemoglobins (e.g. M, Zurich, etc.) are so unstable that they spontaneously oxidise too rapidly for these mechanisms to maintain normal concentrations of methaemoglobin within the cell.

Table II. Causes of methaemoglobinaemia

1. Spontaneous oxidation
2. Agents acting directly as haem group Chlorates Hydrogen peroxide Quinones and quinolines (codeine) Nitrates, nitrites Methylene blue
3. Agents acting mainly *in vivo* Aryl, amino, nitro compounds Aniline Acetophenetidin Nitrobenzene Sulphonamides
4. Unstable haemoglobins (M, Zurich), etc.
5. Methaemoglobin reductase deficiencies

Hereditary methaemoglobinaemia has been shown to be due to a lack of methaemoglobin reductase activity [GIBSON, 2; SCOTT and GRIFFITHS, 9; WEST *et al.*, 10; KAPLAN and BEUTLER, 4], and WEST has recently studied three families with this condition, attempting to define the type of protein defect present by starch gel electrophoresis of enzyme concentrates. The defect could result from a deficiency of the normal enzymes protein, the presence of an abnormal enzyme protein, or one which is unstable. Electrophoresis of enzyme concentrates from affected members of these families have revealed three different enzyme proteins, two being of the same mobility but with different K_M for NADH. The presence of both normal and abnormal enzyme bands was found in heterozygotes. These proteins had a different electrophoretic mobility from the California variant reported by KAPLAN and BEUTLER [4]. Thus hereditary methaemoglobinaemia appears to be due to the synthesis of an abnormal form of enzyme in these patients. In one other family an unstable form of enzyme has been reported [KEITT *et al.*, 5] but it is not yet known whether the enzyme variants in the families reported here are unstable or not.

References

1. DE LOECKER, W. C. J. and PRANKERD, T. A. J.: Factors influencing the hexose monophosphate shunt in red cells. Clin. chem. Act. *6:* 641 (1961).
2. GIBSON, Q. H.: The reduction of methaemoglobin in red blood cells and studies on the cause of idiopathic methaemoglobinaemia. Biochem J. *42:* 13 (1948).
3. HUENNEKENS, F. M.; KERWAR, G. K. and KASITA, A.: Methaemoglobin reductoses. Hereditary disorders of erythrocyte metabolism, p. 87 (Grune and Stratton, New York 1968).
4. KAPLAN, J. C. and BEUTLER, E.: Electrophoresis of red cell NADH and NADPH diaphorase in normal subjects and patients with congenital methaemoglobinaemia. Biochem. biophys. Res. Comm. *29:* 605 (1967).
5. KEITT, A. S.; SMITH, T. W. and JANDL, J. H.: Red cell pseudomosaicism in congenital methaemoglobinaemia. N. Engl. J. Med. *275:* 397 (1966).
6. KIESE, M.: Die Reduktion des Hämoglobins. Biochem. Z. *316:* 264 (1944).
7. SCOTT, E. M. and GRIFFITH, I. V.: The enzyme defect of hereditary methaemoglobinaemia. Biochem. biophys. Acta *34:* 584 (1959).
8. SCOTT, E. M. and MCGRAW J. C.: Purification and properties of diphosphopyridine nuleotide diaphorase of human erythrocytes. J. biol. Chem. *237:* 249 (1962).
9. WEST, C. A.: Ph. D. Thesis, University of London (1969).
10. WEST, C. A.; GOMPERTS, B. D.; HUEHNS, E. R.; KESSEL, I. and TASHBY, J. R.: Demonstration of an enzyme variant in a case of congenital methaemoglobinaemia. Brit. med. J. 212 (1969).

Author's address: Prof. T. A. J. PRANKERD, University College Hospital, Gower Street, *London W.C. 1* (England)

7th int. Congr. clin. Chem., Geneva/Evian 1969; vol. 2: Clinical Enzymology, pp. 23–31
(Karger, Basel/München/Paris/New York 1970)

Glucose-6-Phosphate Dehydrogenase Variants and Its Clinical Implications

B. RAMOT

Department of Hematology, Government Hospital Tel-Hashomer and Tel-Aviv University Medical School

The deficiency in glucose-6-phosphate dehydrogenase (G-6-PD) is probably the most prevalent genetic mutation known at present. It is roughly estimated that about 100 million people are affected by it all over the world.

The mutations of the G-6-PD locus are transmitted on the X-chromosome. The hemizygous male expresses the defect fully while in the heterozygous female the phenotypic expression was found to be variable [1]. This is probably related to the random inactivation of one X-chromosome in every cell as suggested by BEUTLER, using LYON's hypothesis [2, 3].

The clinical manifestations of G-6-PD deficiency differ in various populations. In Negroe mutants the drug-induced hemolytic events were found to be self limited, while in the Mediterranean mutants this was not the case [4, 5].

Furthermore, hemolysis during infections, favism and hemolytic jaundice in the newborn were observed in the latter group [6].

Heterogeneity of the G-6-PD mutations in various populations was also suggested from enzyme level determinations in tissues other than the red cells [7, 8, 9].

In 1962 BOYER and WEILBACHER demonstrated the first qualitatively abnormal G-6-PD [10]. The latter was found in about 18% of normal American Negroe males and is readily distinguished from the wild type Gd B+ by its rapid electrophoretic mobility at alkaline pH [10]. Recently, YOSHIDA has shown that the difference between the two variants is due to one amino acid substitution asparagine to aspartic acid in the A+ variant [11].

Studies by KIRKMAN *et al.* and later by RAMOT *et al.* indicated that the common variant in the Mediterranean population (Gd Mediterranean) is kinetically different from the wild enzyme (Gd B+) [12, 13]. Since new common and rare variants were detected every week or month, a WHO

Table I. Criteria for the identification of G6PD variants

1. Activity in hemolysate
2. Michaelis Menten constant (K_M)
3. Thermal stability
4. Affinity for 2-deoxy G-6-P and other structural analogues
5. Electrophoretic mobility
6. pH curve
7. Stability to SH-inhibitors

standardization committee has established criteria for the identification of new variants, [14] (Table I).

Till now about 50 common and rare variants have been described. They can be grossly classified into 3 groups according to the clinical syndrome they are associated with: a) variants with normal or slightly decreased enzyme activity detected when starch gel electrophoresis was used as a screening method (table II). These variants are of no clinical significance known at present. b) Common and rare variants with decreased or very low enzyme activity (table III). Each of the mentioned variants is prevalent in same populations [14]. The Gd A− was found in about 12% of American Negroes. The affected subjects respond to oxidative drugs by a self limiting intravascular hemolysis. This variant was distinguished chromatographically from the Gd A+ variant [14], indicating a structural difference between the two. This structural mutant is synthesized closely to normal in the young erythrocytes, but its *in vivo* destruction is markedly increased [15]. The normal enzyme levels in the young red cells is the most probable explanation for the self limiting nature of the hemolytic events in the Gd A− variant.

The Mediterranean variant (Gd Mediterranean) is prevalent in the Mediterranean basin and Asia. The hemolytic events are caused not only by oxidative drugs but also during infections and fava bean administration. Contrary to the Gd A− the activity of this variant in young red cells was found to be very low. Using a density separation with phtalate esters we were able to show that the 1% lightest (youngest) red cells have very low enzyme levels and that the rate of enzyme decline during aging appears to be similar to that observed in normals (table IV), thus explaining the continuous hemolysis, even when the red cell population is young, if the drug is further administered.

A similar enzyme behaviour was found in 7 individuals after a fava bean induced hemolysis. On the other hand PIOMELLI, using a cytochemical method,

Table II. Variants with normal or slightly decreased enzyme activity

Variant	RBC activity (% of normal)	Electrophoretic mobility (% of Gd B+)	K_M for G-6-P (μM)	2-d-G-6-P utilization	Heat stability	Optimal pH
King County	100	Normal 100	50–78	<4	Normal	Normal
Madison	100	Slow 70–90	?	?	?	?
Ijebu-Ode	100	Slow	60	?	Decreased	Biphasic
Ita-Bale	100	Slow	91	?	Slightly decreased	Normal
Baltimore-Austin	75	Slow 90	68	<4	Normal	Normal
Ibadan-Austin	72	Slow 80	62–72	<4	Normal	Normal
Minas Gerais	<70	95	41	9	?	Normal
Madrona	70–80	80	32	Normal	?	Normal
Barbieri	40–60	Fast 125	Increased	?	Normal	?
Capetown	53–80	55–65	11–14	7–16	Normal	Biphasic
Kerala	50	Slow 75–90	23	7.4	Normal	Biphasic
Hektoen	400–500	100	?	?	?	?

Table III. Common and rare variants with decreased or very low enzyme activity

Variant	RBC activity (% of normal)	Electrophoretic mobility (% of Gd B+)	K_M for G-6-P (μM)	2-d-6-G-P utilization	Heat stability	Optimal pH
Mediterranean	0.7	Normal	19–26	25–40	Low	Biphasic
A-	8–20	Fast 110	Normal	<5	Normal	Normal
Canton	4–24	Fast 105	20–36	4–15	Slightly reduced	Biphasic
Markham	15–10	105–108	4.4–6.3	162–222	Reduced (?)	Very biphasic
Athens	20	98	19	15	Slightly reduced	Slightly biphasic
Columbus	35	Normal	Normal	Normal	?	?
Puerto Rico	19	112	18.6	2.7	Slightly decreased	Normal (?)
Tel-Hashomer	25–40	Slow 67–70	30–40	Normal	Normal	Slightly biphasic
Washington	16	95	57	1.6	Normal	Normal
Kabyle	14–36	104 (TEB) 110 (phosphate)	68	10	Normal	Normal
West-Bengal	9	Slow 90	31	4	Normal	Normal
Seattle	8–21	Slow 90	15–25	7–11	Normal	Slightly biphasic
Hong-Kong	<8	100	58	5–4	Markedly diminished	7.0
Benevento	7	93	4.6	245	Decreased	5.5; 9.75
Panay	<5	96	30	Normal	Slightly increased	Biphasic
Union	<3	Fast	8–12	180	?	Biphasic
Kephalonia	Low	Fast	34.2	Increased	Normal	Truncate
Attica	Low	Slow	40	Increased	Normal	Truncate
'Seattle Like'	8–21	Slow 90	15–25	7–11	Normal	Biphasic
Lifta	0	90	25	60	Very labile	Abnormal

Table IV. Glucose-6-phosphate dehydrogenase activity[1] in oldest and youngest cells in normal, Mediterranean mutants and during favism

Case	NO. OF exp.	1% 'Old' cells		2% 'Old' cells		Total cell population		2% 'Young' cells		1% 'Young' cells	
		Mean	S.D.	Mean	S.D.	Mean	S.D.	Mean	S.D.	Mean	S.D.
Normal	52	1.30	0.40	2.00	0.50	2.50	0.60	3.20	0.30	4.30	0.30
G-6-PD deficiency	20	0.05	0.03	0.10	0.07	0.19	0.10	0.30	0.20	0.58	0.23
Favism	7	0.28	0.20	0.33	0.30	0.42	0.30	0.62	0.40	0.96	0.60

[1] Expressed as μmoles/min/gr Hb.

found normal enzyme activity in the bone marrow erythroblasts of such individuals. His interpretation for the low enzyme activity is *in vivo* destruction of the enzyme in the normoblats and reticulocytes as a result of marked enzyme lability. The erythroblast G-6-PD levels need further investigation.

The Canton variant (Gd Canton) appears to be very similar to the Mediterranean variant [16]. The clinical syndromes associated with this variant are not fully described. Favism is probably not rare among them.

The Markham variant (Gd Markham) found in New Guinea is kinetically very interesting, however, the clinical syndromes associated with this variant are not well documented [17].

To make gross generalizations it may be said that the clinical manifestations of the second group of variants is at least partly related to the enzyme level. A variant with 30–50% enzyme activity is rarely associated with severe drug induced hemolytic episodes and they rarely if ever hemolyse during infections. The latter point should be further investigated under well controlled conditions.

Uncommon variants with low or very low enzyme activity were isolated from cases of congenital non-spherocytic hemolytic disease (CNHD). Nineteen such variants were characterized (table V) and more are probably on the way. During the last two years we have studied 5 patients with indirect hyperbilirubinemia and G-6-PD deficiency [18]. All of them had severe hemolytic episodes following infections, drug administration and frequently without a demonstrable cause. The hemoglobin levels in 4 of the patients were normal. However, all had a slightly increased reticulocyte count and an indirect hyperbilirubinemia. Three new variants were isolated and characterized by the methods specified by the WHO [14]. The pH curves were performed by the method of KIRKMAN [17]. Two of the patients had the common Mediterranean variant, in spite of the clinical manifestations of a non-spherocytic hemolytic disease. A number of such cases have been described previously [19].

The common features characteristic to this group are increased lability of the paritially purified enzyme *in vitro* and the high K_M for glucose-6-phosphate.

There is no explanation at present for the chronic hemolysis associated with G-6-PD deficiency. This is particularly unclear in patients in whom the Mediterranean variant was found. One explanation, suggested by BEUTLER, was a second genetic mutation not determined at present [19]. Another possibility is that the Gd Mediterranean variants in cases of CNHD differ from the common Mediterranean variants but this difference cannot be

Table V. Uncommon variants with low or very low enzyme activity associated with congenital non-spherocytic hemolytic disease

Variant	RBC activity (% of norm.)	Electrophoretic mobility (% of Gd B+)	K_M for G-6-P µM	2-d-G-6-P utilization	Heat stability	Optimal pH
Chicago	9–26	Normal	58–76	<4	Very labile	Normal
Duarte	8.5	100	58	5.4	Very labile	7.0
Freiburg	10–20	Slow 85–90	87–118	?	?	Biphasic
Oklahoma	4–10	Normal	127–200	<4	Decreased	Narrowpeak
Paris	4	?	280	?	Very labile	Peak 9.5
Ohio	2–16	Fast 110	Slightly increased	Normal	Very labile	?
Torrance	2.4	103 (phosphate)	48–60	2.4	Very labile	Normal
Clichy	2	100	178	?	?	Abnormal plateau 9–10
Albuquerque	1	100	115	0	Very labile	8.5
Milwaukee	0.5	92	224	3.7	?	8
Tübingen	0.3	100	Decreased	?	?	?
Berlin	0–1	?	Increased	0	?	?
Eyssen	0	Slow 90	?	?	Decreased	?
Beaujon	0	Fast	182	?	?	Peak 9.5
Ramat-Gan	0	92	35	40	Very labile	Biphasic 6.5;10
Bat-Yam	0	100	27	45	Very labile	Biphasic 6.5;10
Ashdod	10	92	100	40	Stable	Biphasic 6.5;10
Mediterranean	0.7	Normal	19–26	25–40	Decreased	Biphasic

detected with the present techniques. Such examples are well known from the field of abnormal hemoglobins.

In spite of the large amount of information that has accumulated the unknown is much greater than the known. Methods for purification and fingerprinting of G6PD from small amounts of blood will advance this fast moving field.

Summary

G-6-PD variants are divided into three major groups:

1) Variants with normal or slightly decreased enzyme activity detected when population screening is performed.
2) Variants with low enzyme activity associated with a drug induced hemolytic process or anemia.
3) Uncommon variants with low enzyme activity observed in association with non spherocytic hemolytic anemia or hemolytic process.

The criteria for their characterization and the correlation between some kinetic properties of the enzyme and the clinical manifestations is discussed.

References

1. BEUTLER, E.: G-6-PD deficiency; in J. B. STANBURY, J. B. WYNGAARDEN and O. S. FREDRICKSON: The metabolic basis of inherited diseases, p. 1,061 (Mc Graw-Hill, New York 1965).
2. LYON, M. F.: Gene action in the X-chromosome of the mouse. Nature, Lond. *19:* 372 (1961).
3. BEUTLER, E.; YEH, M. and FAIRBANKS, F.: The normal human female as mosaic of X-chromosome activity. Studies using the gene for G-6-PD deficiency as a marker. Proc. nat. Acad. Sci. *48:* 9 (1962).
4. DERN, R. S.; BEUTLER, E. and ALVING, A. S.: The hemolytic effect of primaquine. II. The natural course of the hemolytic anemia and the mechanism of its self limited character. J. Lab. and clin. Med. *44:* 171 (1954).
5. SZEINBERG, A.; SHEBA, C. and ADAM, A.: Enzymatic abnormality in erythrocytes of a population sensitive to *Vicia fava* or haemolytic anemia induced by drugs. Science *124:* 484 (1956).
6. FESSAS, P.; DOXIADES, S. A. and VALAS, T.: Neonatal jaundice in G-6-PD deficient infants. Brit. med. J. *ii:* 1,359 (1962).
7. MARKS, P. A. and GROSS, R. T.: Erythrocyte glucose-6-dehydrogenase deficiency: Evidence of difference between Negroes and Caucasians with respect to this genetically determined trait. J. clin. Invest. *38:* 2,253 (1959).

8. RAMOT, B.; SHEBA, Ch.; SZEINBERG, A.; ADAM, A.; ASHKENAZI, L. and FISHER, S.: Further investigation of erythrocyte G-6-PD deficient subjects. Enzyme levels in other tissues and its genetic implication. Proc. Int. Congr. Haemat., Tokio 1960.
9. RAMOT, B.; FISHER, S.; SZEINBERG, A.; ADAM, A.; SHEBA, Ch. and GAFNI, D.: A study of subjects with erythrocyte G-6-PD deficiency: Investigation of platelet enzymes. J. clin. Invest. *38:* 1,659 (1959).
10. BOYER, S. H.; PORTER, I. H. and WEILBACHER, R. G.: Electrophoretic heterogeneity of G-6-PD and its relationship to enzyme deficiency in man. Proc. nat. Acad. Sci. *48:* 1,868 (1962).
11. YOSHIDA, A.: A single amino acid substitution (asparagine to aspartic acid) between normal (B+) and the common Negro variant (A+) of human G-6-PD. Proc. nat. Acad. Sci. *57:* 835 (1967).
12. KIRKMAN, H. N.; SCHETTINI, F. and PICKARD, B. M.: Mediterranean variant of glucose-6-phosphate dehydrogenase. J. Lab. and clin. Med. *63:* 726 (1964).
13. RAMOT, B.; BAUMINGER, S.; BROK, F.; GAFNI, D. and SHWARZ, Y.: Characterization of glucose-6-phosphate dehydrogenase in Jewish mutants. J. Lab. clin. Med. *64:* 895 (1964).
14. World Health Organization Technical Report Series No. 366: Standardization of Procedures for the Study of G-6-PD. Report of A WHO Scientific Group (1967).
15. YOSHIDA, A.; STAMATOYANNOPOULOS, G. and MOTULSKY, A. G.: Negro variants of G-6-PD deficiency (A—) in man. Science *155:* 97 (1967).
16. MCCURDY, P. R.; KIRKMAN, H. N.; NAIMAN, J. L.; JIM, R. T. S. and PICKAR, B. M.: A Chinese variant of G-6-PD. J. Lab. clin. Med. *67:* 374 (1968).
17. KIRKMAN, H. N.; KIDSON, C. and KENNEDY, M.: Variants of human G-6-PD. Studies of samples from New-Guinea; in E. BEUTLER, Hereditary disorders of erythrocyte metabolism, p. 126. City of Hope Symposium Series, vol. 1 (Grune and Stratton, New York 1968).
18. RAMOT, B.; BEN-BASSAT, I. and SHCHORY, M.: New glucose-6-phosphate dehydrogenase variants observed in Israel. Association with congenital non-spherocytic hemolytic disease. J. Lab. clin. Med. (in Press) (1969).
19. BEUTLER, E.; MATHAI, C. K. and SMITH, J. E.: Biochemical variants of G-6-PD giving rise to congenital nonspherocytic hemolytic disease. Blood *31:* 131 (1968).

Author's address: Dr. BRACHA RAMOT, Department of Hematology, Government Hospital Tel-Hashomer and Tel-Aviv University Medical School, *Tel-Aviv* (Israel)

7th int. Congr. clin. Chem., Geneva/Evian 1969; vol. 2: Clinical Enzymology, pp. 32–39 (Karger, Basel/München/Paris/New York 1970)

Proteaseninhibitoren — Bedeutung für Forschung und Klinik

E. WERLE

Institut für Klinische Chemie und Klinische Biochemie der Universität München, München (Direktor: Prof. Dr. Dr. E. WERLE)

Eine Reihe von physiologisch und pathologisch bedeutsamen Vorgängen im Organismus ist mit einer « Kaskade » von proteolytischen Mechanismen verbunden, so die Blutgerinnung, die Fibrinolyse, die Bereitstellung von Verdauungsenzymen des Pankreas, die Mobilisierung der Plasmakinine und des Angiotensins II.

Auch die Freisetzung einiger Hormone, z.B. des Insulins und des Schilddrüsenhormons, erfolgt durch Proteolyse. Fast alle in diesen Prozessen wirksamen Enzyme liegen in den Geweben bzw. im Blutplasma als inaktive Vorstufen vor und werden durch begrenzte Proteolyse, also durch Lösen einer oder doch nur sehr weniger Peptidbindungen in den großen Molekülen in die aktiven Formen übergeführt.

Es handelt sich bei all diesen Mechanismen um Möglichkeiten zur Steuerung oder Dosierung funktionell wichtiger Vorgänge. Eine weitere Einrichtung zu ihrer Kontrolle steht dem Organismus in Form von Proteaseninhibitoren zur Verfügung [VOGEL *et al.*, 10; WERLE, 12].

In der Tabelle I sind nur die relativ niedermolekularen natürlichen Inhibitoren mit ihrem Hemmspektrum aufgeführt. Ihre physiologische Funktion ist in allen Fällen noch Gegenstand der Forschung. Daneben gibt es im Blutplasma eine Reihe hochmolekularer Inhibitoren von Eiweißnatur.

Physiologischerweise tragen diese Inhibitoren dazu bei, die angedeuteten Prozesse in den für die Bedürfnisse des Organismus erwünschten Grenzen zu halten. Es gibt aber zahlreiche pathologische Vorgänge, bei denen diese Schutzeinrichtung durchbrochen wird.

Ein Beispiel: Die meisten Verdauungsenzyme des Pankreas sind in Form von Proenzymen in den Acini gespeichert und werden als solche in den Pan-

Tabelle I

Inhibitor	Trypsin O = temporär[1]	Chymo.	Plasmin	Kallikrein Pankreas	Kallikrein Serum	Thrombin	Mol. Gew. bei
Trasylol (Rind)	+	+	+	+	+	+	6,500
Pankreas (Säuger)	⊕	—	—	—	—	—	6,000
Samenblasen	⊕	—	+	—	—	—	6,000
Gl. submand.	⊕	+	—	—	—	—	6,000
Bronchialsekret	—	+		—			H^+-stabil
Hirudin	—	—	—	—	—	+	9,000
'Bdellin'	+	(+)	+	—	—	—	6,000
Sojabohnen	+	+	+	—	+	—	22,000
Eiklar	⊕	—	—	—	+	—	26,000

[1] Auch Abbau durch Pepsin!

kreassaft abgegeben. Durch die Enterokinase des Darmsaftes wird zuerst Trypsinogen in Trypsin umgewandelt. Dieses aktiviert dann durch begrenzte Proteolyse Chymotrypsin, Carboxypeptidase, Kallikrein und, wie man seit kurzem weiß, auch Phospholipase A [WANKE *et al.*, 11] aus ihren Vorstufen. Gegen das vorzeitige Aktivwerden dieser Enzyme enthält das Pankreas einen spezifischen Inhibitor für Trypsin. Er könnte 5–10% des aus Trypsinogen maximal aktivierbaren Trypsins blockieren, was physiologischerweise ausreicht, um die intraglanduläre Aktivierung der Enzymkette zu verhindern. Für aktives Kallikrein besitzt das Pankreas, mit Ausnahme das der Rinder, keinen wirksamen Inhibitor [FREY *et al.*, 1; VOGEL *et al.*, 10].

Wird nun aufgrund pathologischer Prozesse Trypsin in größerem Maß schon intraglandulär aktiviert, so wird die Inhibitor-Barriere durchbrochen und es kommt zum Erscheinungsbild der Pankreatitis. Dabei spielt u.a. die in der ersten Abbildung wiedergegebene Aktivierung des Kallikreins eine bedeutsame Rolle (Abb. 1). Das aktive Kallikrein liberiert aus dem Kallidinogen der interstitiellen Gewebsflüssigkeit durch begrenzte Proteolyse Kinine, das sind pharmakologisch hochaktive Polypeptide, die folgende Wirkungen haben: Erhöhung der Kapillarpermeabilität, die Ödembildung verursachen kann, Gefäßerweiterung, Blutdrucksenkung, Schmerzerzeugung. Sie können also alle Erscheinungen von Entzündung auslösen [FREY *et al.*, 1]. Man kann 3 Kinine unterscheiden, wie Abbildung 2 zeigt.

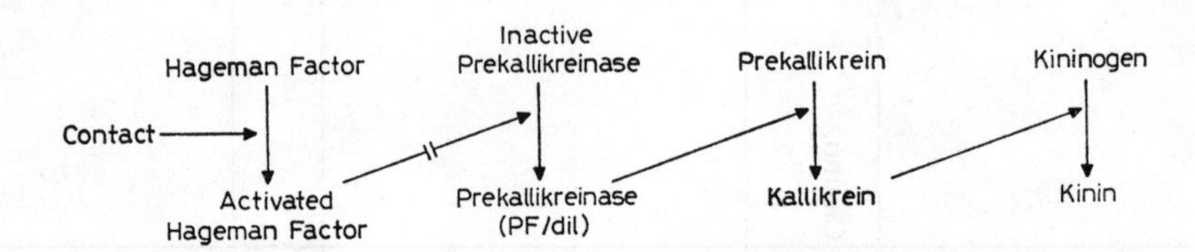

Abb. 1. Mechanismus der Aktivierung von menschlichem Plasmakallikrein [WEBSTER, M. E. und INNERFIELD, I.; Enzymol. biol. clin. *5:* 129, 1965].

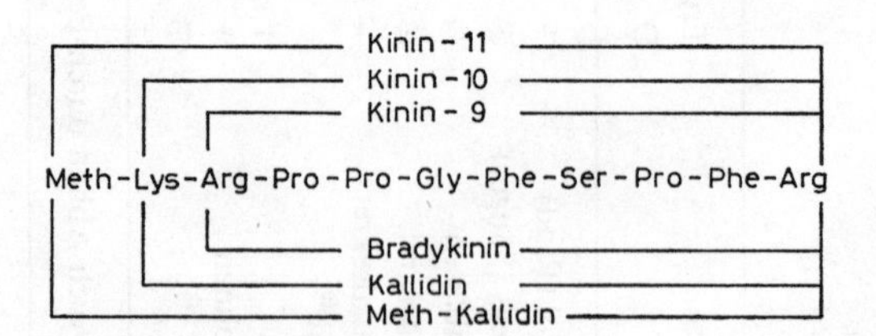

Abb. 2. Aminosäuresequenz der Plasmakinine.

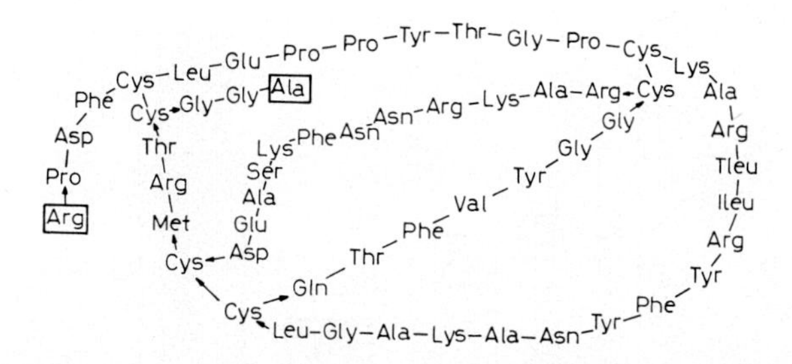

Abb. 3. Aminosäuresequenz des Kallikrein-Trypsin-Inhibitors nach KASSELL und LASKOWSKI [KASSEL, B. und LASKOWSKI, M.: Biochem. biophys. Res. Commun. *20:* 463, 1965] und ANDERER [ANDERER, F. A.: Z. Naturforsch. *20b:* 462, 1965].

Man versucht nun durch Verabreichung des polyvalenten Inhibitors aus Rinderorganen, also des Trasylol® (Abb. 3), diese Reaktionskette zu unterbrechen, die u.a. zum Schock führen kann.

Trasylol besitzt unter allen Inhibitoren das breiteste Hemmspektrum: Es hemmt außer Trypsin auch Chymotrypsin, Plasmin und Plasmin-Aktivatoren, ferner Proteasen des Gerinnungssystems, Organ- und Plasmakallikreine und schließlich leukozytäre und lysosomale Proteasen [VOGEL *et al.*, 10]. In jedem Fall handelt es sich um Serinenzyme, so genannt, weil die Blockierung des Serinrestes im aktiven Zentrum der Enzyme, z.B. durch Phosphorylierung mit DFP, die Wirkung aufhebt. Eine Ausnahme stellt der Hageman-Faktor dar, der obwohl er zu den Serinenzymen und zu den Peptid-peptido-Esterasen gehört, durch Trasylol nicht gehemmt wird [HOCHSTRASSER und WERLE, 4].

Im Gegensatz zu den hochmolekularen Inhibitoren des Blutplasmas, unter denen auch Kallikrein-Inhibitoren sich befinden, vermag Trasylol leicht zu permeieren und an die Orte pathologischen Geschehens in den Geweben zu gelangen. Es verschwindet relativ rasch aus dem Blut und wird schließlich in der Niere angereichert [REICHENBACH-KLINKE *et al.*, 9]. Nur 1,5% davon werden im Harn in aktiver Form nachweisbar [HAENDLE und WERLE, 3] (Abb. 4).

Dabei wird das Trasylol in der Niere vorübergehend an einen hochmolekularen Eiweißkörper, möglicherweise an ein Enzym, gebunden [REICHENBACH-KLINKE *et al.*, 9]. Aus Niere in niedermolekularer Form zurückgewonnenes Trasylol ist durch das Fehlen der N- und C-terminalen Aminosäure (s. Abb. 3) des intakten Trasylolmoleküls, also von Arginin bzw. Alanin, gekennzeichnet. Dieser verkürzte Inhibitor ist in bezug auf seine Hemm-

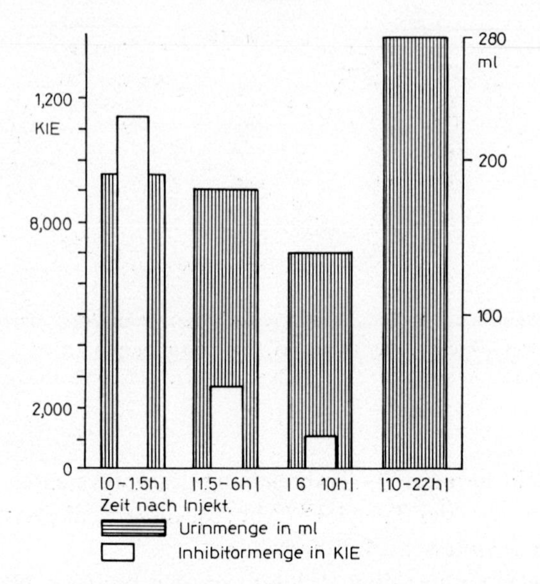

Abb. 4. Ausscheidung von Trasylol nach intravenöser Injektion von 1 Mill. KIE beim Menschen.

kapazität, z.B. für Trypsin, unverändert. Die Fixierung des Trasylol im Nierenparenchym ist auf seine hohe Basizität zurückzuführen [REICHENBACH-KLINKE *et al.*, 9]. Wird diese abgeschwächt, z.B. durch Bildung von Tetramaleoyl-Trasylol, so ist seine Eliminierung aus dem Blut verlangsamt und die Substanz wird nahezu quantitativ in den Harn ausgeschieden [WERLE und FRITZ, 13].

In ähnlicher Weise wie bei der Pankreatitis könnte Trasylol bei allen pathologischen Prozessen eingesetzt werden, bei denen aufgrund einer Aktivierung des Plasma-Kallikreins die erwähnten Kinine liberiert werden. So beim Carcinoid, bei Arthritis, beim dumping syndrom und anderen pathologischen Prozessen.

Die ödemhemmende Wirkung von Proteaseninhibitoren wurde am Beispiel des Pfotenödems der Ratte von KALLER, [5] und von WERLE und TAEGER, [14] demonstriert. Die pflanzlichen Inhibitoren hemmen an anderer Stelle wie das Trasylol. Während das Trasylol aktiv gewordenes Plasmakallikrein hemmt, verhindern oder hemmen die pflanzlichen Inhibitoren, soweit untersucht, die Überführung von Plasmapräkallikrein in aktives Plasmakallikrein. Nach orientierenden Versuchen ist hier Angriffspunkt der Hageman-Faktor, da die Inhibitoren aus Bohnenarten die Esterasewirkung

(BAEE) des Hageman-Faktors stark inhibieren können [HOCHSTRASSER und WERLE, 4].

Aufgrund seines breiten Hemmspektrums vermag Trasylol, wie erwähnt, die Reaktionskette, die zur Blutgerinnung führt, sowie die Vorgänge der Fibrinogenolyse und Fibrinolyse zu inhibieren. Die Enzyme, die dabei gehemmt werden, sind in den nächsten Abbildungen (Abb. 5 und 6) aufgezeichnet.

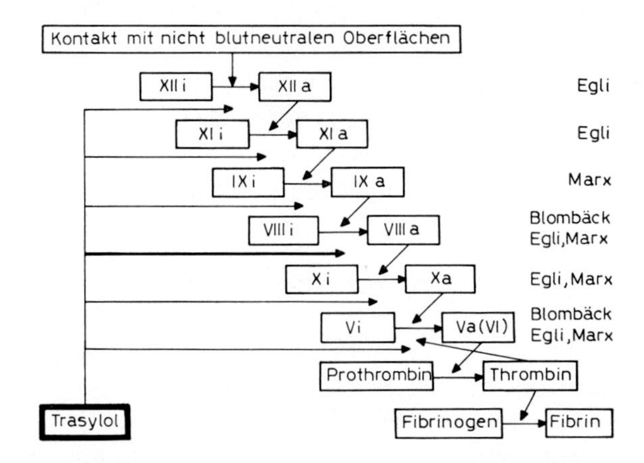

Abb. 5. Hemmung der pathologisch gesteigerten Blutgerinnungsfähigkeit durch Trasylol [nach: HABERLAND, H. G. und MATIS, P.: Trasylol ein Proteinase-Inhibitor bei chirurgischen und internen Indikationen. Med. Welt *18:* 1,367–1,376, 1967].

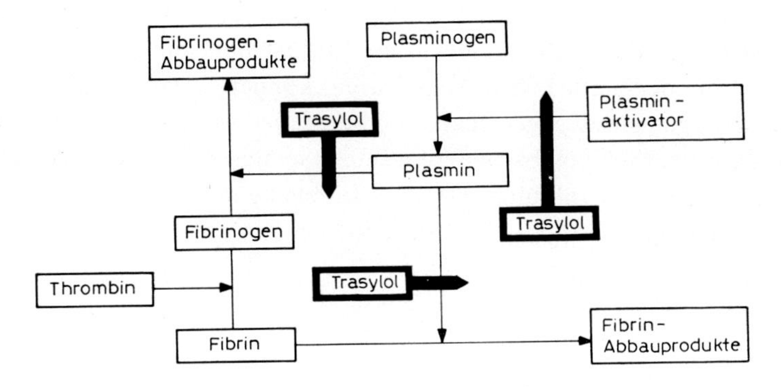

Abb. 6. Hemmung der Fibrinolyse und Fibrinogenolyse durch Trasylol [nach: HABERLAND, H. G. und MATIS, P.: Trasylol ein Proteinase-Inhibitor bei chirurgischen und internen Indikationen. Med. Welt *18:* 1,367–1,376, 1967].

Der einzige bislang für Thrombin aufgefundene natürliche niedermolekulare Hemmstoff wurde in den Arbeitskreisen von MARKWARDT [8] und DE LA LLOSA *et al.* [7] aus Blutegelextrakten isoliert. Die gerinnungshemmende Wirkung von Blutegelextrakten ist zwar schon lange bekannt und wurde auch therapeutisch genützt, eine breitere Anwendung blieb diesen Hirudinpräparaten jedoch versagt. Der Grund dafür dürfte darin zu suchen sein, daß die Hirudinpräparate neben dem Thrombin-Inhibitor noch Plasmin-Inhibitoren in hoher Konzentration (sie machen ca. 20% der Präparate aus) enthielten. Es wurden also therapeutisch sich ausschließende Inhibitorgemische verwendet [WERLE *et al.*, 13]. Die von uns aus Blutegelextrakten isolierten Plasmin-Inhibitoren besitzen eine ähnlich starke Antiplasminwirkung wie Trasylol; ihr Molekulargewicht liegt mit ca. 5000 noch niedriger als das des Trasylol und des spezifischen Pankreas-Trypsin-Inhibitors. Sie werden nach i.v. Injektion, ebenso wie der Thrombinhemmstoff, mit dem Harn ausgeschieden.

H.Pyroglu-Lys-Trp-Ala-Pro.OH.

Abb. 7

Über eine neue Inhibitorklasse mit vielversprechenden Eigenschaften haben kürzlich GREENE *et al.* [2], und KATO und SUZUKI [6] berichtet. GREENE isolierte aus dem Gift von *Bothrops jararaca* ein hemmendes Pentapeptid (Abb. 7), das er auch synthetisieren konnte. Es hemmt Kininasen in äußerst geringer Konzentration, insbesondere die Kininase der Lunge. Bei den von KATO und SUZUKI aus dem Gift von *Agkistrodon blomhofii* isolierten Polypeptid mit 8–10 Aminosäuren mit ähnlich starker Kininase-Hemmwirkung fällt besonders der hohe Prolingehalt auf.

Mit Hilfe der natürlichen Inhibitoren können viele Proteinasen des tierischen Organismus gegeneinander abgegrenzt werden. Darüber hinaus sind sie therapeutisch von besonderem Interesse. Besonders wichtig erscheint, daß das Studium der Inhibitoren tiefere Einblicke in die Wechselwirkungen und gegenseitigen Bindungen von Proteinen eröffnet.

Summary

In plants and animals there occur inhibitors for proteases especially for trypsin, chymotrypsin and plasmin: they are polypeptides. Those of the animal body have molecular weights of around 6000. The seeds of plants like soya beans contain such inhibitors with molecular weights of 8000 to 22,000 and blood plasma with 60,000. The physiological function of protease inhibitors in the animal body is presumably to hold proteolytic processes in physiological border. Many proteolytic enzymes of the animal body may be differentiated by means of such inhibitors. Study of the basic chemistry of the inhibitors, their kinetic action and active sites provides new insight into possible mechanisms of the inhibitor enzyme binding. Some of the inhibitors are of therapeutical interest.

Literatur

1. FREY, E. K.; KRAUT, H.; WERLE, E.; VOGEL, R.; ZICKGRAF, G. und TRAUTSCHOLD, I.: Das Kallikrein-Kinin-System und seine Inhibitoren (Enke, Stuttgart 1968).
2. GREENE, L. J.; STEWART, J. M. and FERREIRA, S. H.: Bradykininpotentiating peptides from the venom of *Bothrops jararaca.* Pharmacol. Res. Commun. *1:* 159–160 (1969).
3. HAENDLE, H. und WERLE, E.: unveröffentlicht (1969).
4. HOCHSTRASSER, K. und WERLE, E.: unveröffentlicht (1969).
5. KALLER, H.: Die Beeinflussung experimentell erzeugter Ödeme durch Trasylol; in R. GROSS und G. KRONEBERG Neue Aspekte der Trasylol Therapie, pp. 152–158 (Schattauer, Stuttgart 1966).
6. KATO, H. and SUZUKI, T.: Isolation and identification of bradykinin potentiators in snake venoms. Pharmacol. Res. Commun. *1:* 166–167 (1969).
7. DE LA LLOSA, P.; TERTRIN, C. et JUTISZ, M.: L'enchaînement C-terminal de l'hirudine. Biochim. biophys. Acta *93:* 40–44 (1964).
8. MARKWARDT, F.: Blutgerinnungshemmende Wirkstoffe aus blutsaugenden Tieren (Fischer, Jena 1963).
9. REICHENBACH-KLINKE, K. E.; MECKL, D.; KEMKES, B.; HOCHSTRASSER, K.; FRITZ, H. und WERLE, E.: Die Fixierung und Veränderung eines Proteasehemmstoffes in der Niere. Arzneimittelforsch. *19:* 1,025–1,026 (1969).
10. VOGEL, R.; WERLE, E. and TRAUTSCHOLD, I: Natural proteinase inhibitors (Academic Press, New York 1968).
11. WANKE, M.; NAGEL, W.; LINDER, M. M. und SEBENING, H.: Über die Stellung der Phospholipase A im Ablauf der akuten Pankreatitis. Gastroenterolologie *6:* 434–442 (1968).
12. WERLE, E.: Plasmakinine. Dtsch. med. Wschr. *92:* 1,573–1,580 (1967).
13. WERLE, E. und FRITZ, H.: unveröffentlicht (1969).
14. WERLE, E. und TAEGER, I.: unveröffentlicht (1969).

Adresse des Autors: Prof. Dr. Dr. E. WERLE, Nussbaumstrasse 20, *D-8 München 15* (Germany)

7th int. Congr. clin. Chem., Geneva/Evian 1969; vol. 2: Clinical Enzymology, pp. 40–45
(Karger, Basel/München/Paris/New York 1970)

Evolution du pouvoir inhibiteur du sang vis-à-vis de certains enzymes au cours de syndromes d'hypercoagulation

E. G. VAIREL et P. FORLOT
Laboratoire Choay

Après avoir montré qu'au cours de pancréatites le sérum des malades donne une ligne de précipitation en immunodiffusion avec des immunsérums préparés à l'aide d'enzymes purs d'origine pancréatique, sachant, d'autre part, qu'au moins l'un d'entre eux, la trypsine, accélère le processus de la coagulation sanguine, nous avons étudié l'évolution des différents paramètres du système coagulo-lytique au cours de pancréatites hémorragiques expérimentales chez le chien.

Nous avons trouvé et rapporté qu'au cours des 90 premières minutes une consommation importante (30 à 40%) et significative des facteurs de coagulation était observée. Dans le même temps, 80% du plasminogène, principal composant du système fibrinolytique, sont consommés. Ceci est un tableau de coagulation intravasculaire disséminée (CIVD) avec réaction fibrinolytique ou coagulopathie de consommation. Cependant, nous n'avons pas trouvé de variation significative de l'hématocrite, du nombre de thrombocytes ni de la coagulation globale représentée par l'indice de potentiel thrombodynamique (IPT) selon RABY. Ceci pourrait, éventuellement, être considéré comme une possibilité de différenciation des origines de CIVD.

La quasi-isocoagulabilité globale reflétée par la thrombélastographie peut s'expliquer par l'augmentation de l'activité plaquettaire sous l'action de la trypsine qui compenserait l'hypofibrinogénémie résultant de la consommation.

En clinique, il est exceptionnel d'être en présence du malade au moment du déclenchement d'une pancréatite et les facteurs consommés ont souvent déjà été remplacés au moment où le prélèvement peut être effectué.

Nous avions remarqué qu'au cours des coagulations intravasculaires disséminées, la consommation des antiplasmines était très importante, mais que, en même temps, on observait une élévation du pouvoir antitrypsine du plasma. Ceci peut provenir d'une stimulation commune au remplacement des deux inhibiteurs, l'abaissement du taux de l'un d'eux entraînant la libération des deux, le taux de celui qui n'était pas consommé s'élève alors au-dessus de la normale.

Nous avons évalué le pouvoir inhibiteur de plasmine et de trypsine du plasma d'un certain nombre de malades atteints de pancréatites, d'infarctus et de CIVD d'origine diverse.

La diversité des cas de CIVD ne nous a pas permis d'obtenir d'analyse statistique satisfaisante et nous n'avons, pour l'instant, que des impressions subjectives.

En revanche, nous avons pu analyser statistiquement les pancréatites et les infarctus; nos résultats sont significatifs.

Sur chaque prélèvement, nous avons fait:

1. une analyse en immunodiffusion contre un sérum antitrypsine;
2. une évaluation du plasminogène;
3. une mesure de l'activité antiplasmine;
4. un dosage du pouvoir inhibiteur de la trypsine.

Les substrats utilisés pour mesurer les activités enzymatiques ont été l'acétyl-glycyl-lysine-méthyl-ester (AGLME) pour la plasmine, le benzoyl-arginine-para-nitro-anilide (BAPNA) pour la trypsine.

Dans le cas des pancréatites, le traitement par les inhibiteurs de protéases ne nous a pas permis de suivre l'évolution, nous avons dû nous satisfaire d'un seul prélèvement effectué avant toute thérapeutique.

Pour chacun des dosages, l'analyse de la variance a été effectuée à l'intérieur des groupes et entre les groupes d'échantillons. Les tests de comparaison ont été calculés pour un risque $\alpha = 0{,}05$.

Le tableau I résume les résultats que nous avons obtenus pour l'évaluation du plasminogène.

Le taux, dans les cas de pancréatite, n'est pas encore restitué totalement; les infarctus sont normaux.

Le tableau II illustre nos résultats dans l'évaluation des antiplasmines; seuls les cas de pancréatites possèdent un pouvoir inhibiteur supérieur aux témoins (127,4%).

Le tableau III indique les valeurs trouvées pour le pouvoir inhibiteur de la trypsine. Les infarctus ne sont pas différents des témoins; les chiffres obtenus pour les pancréatites montrent une augmentation nette (129%). L'intensité de la réaction en immunodiffusion a été notée de 0 à 3 croix.

Tableau I. Dosage du plasminogène dans des plasmas normaux et pathologiques exprimé en UAE/l/min.

Nombre de dosages	Témoins	Pancréatites	Infarctus
	354	92	45
Range	720–2980	540–1960	800–2345
Moyenne ± erreur standard	1500 ± 50	1407 ± 6,1	1512 ± 16
Comparaison par rapport aux témoins	—	significatif	non significatif

Tableau II. Dosage des antiplasmines dans des plasmas normaux et pathologiques en pourcentage d'inhibition de 5 UT/0,1 ml de prise d'essai

Nombre de dosages	Témoins	Pancréatites	Infarctus
	354	92	45
Range	2,5–28,5	11–37	5–28
Moyenne ± erreur standard	14,9 ± 3,7	19,0 ± 2,3	15,7 ± 3,9
Comparaison par rapport aux témoins	—	significatif	non significatif

Tableau III. Dosage des antitrypsines dans des plasmas normaux et pathologiques en pourcentage d'inhibition de 25 μg de trypsine/0,1 ml de prise d'essai

Nombre de dosages	Témoins	Pancréatites	Infarctus
	354	92	45
Range	20–62	27–77	22–55
Moyenne ± erreur standard	39,5 ± 1,2	51,0 ± 2,0	41,8 ± 7,2
Comparaison par rapport aux témoins	—	significatif	non significatif

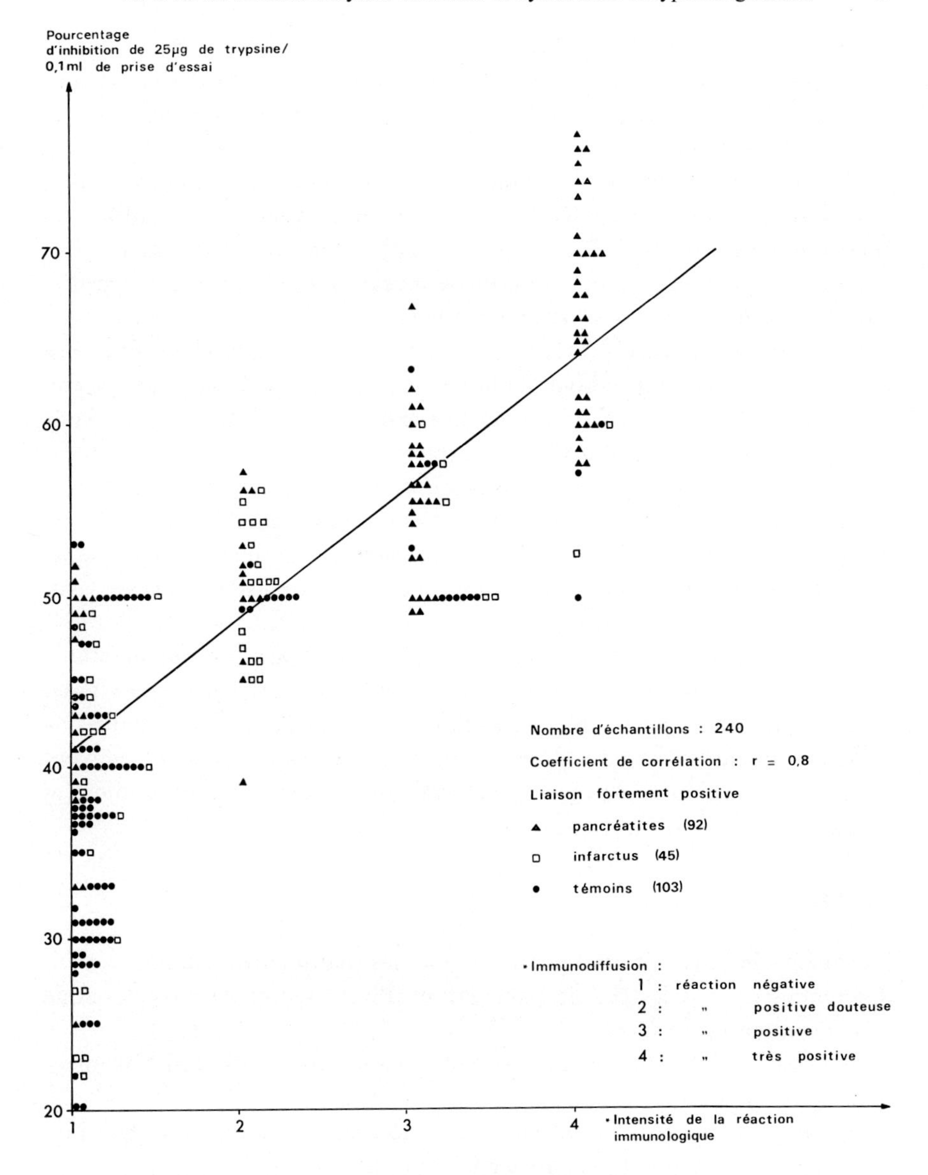

Fig. 1. Etude de la corrélation entre le dosage enzymatique des antitrypsines et l'immunodiffusion contre un sérum antitrypsine.

La figure 1 représente une étude de la corrélation entre le dosage des antitrypsines et l'intensité de la réaction immunologique.

On peut conclure que les paramètres considérés varient de façon significative pour les plasmas provenant de pancréatites par rapport aux plasmas témoins, alors que cette variation n'est pas significative pour les plasmas provenant d'infarctus.

En outre, quelle que soit l'origine des plasmas, une liaison très forte entre le dosage des antitrypsines et le test en immunodiffusion a été observée. Il est intéressant de noter à ce propos la répartition des points par rapport à la droite de régression, qui confirme le niveau plus élevé des antitrypsines dans les plasmas provenant de pancréatites.

Il est certain que les CIVD sont d'origines très diverses, de même peut-être que les syndromes thrombo-emboliques. Le nombre de cas pour chacun des groupes doit être important pour une exploitation mathématique. Nous ne pouvons aujourd'hui fournir cette exploitation que pour les pancréatites; les autres cas ne donnent lieu qu'à des impressions qui sont les suivantes:

1. L'acte chirurgical déclenche des CIVD intenses en chirurgie sous circulation extra-corporelle, abdominale profonde et vasculaire par exemple. Nous y notons une baisse du plasminogène, des antiplasmines et, souvent, une augmentation des antitrypsines.

2. Les insuffisants rénaux ont très fréquemment un tableau identique qui nous laisserait supposer des micro-coagulations provenant de CIVD.

3. Les infarctus ne semblent pas montrer de variations exploitables.

4. Les quelques cas d'embolies que nous avons pu étudier semblent montrer des variations du type CIVD avec diminution des antiplasmines et augmentation des antitrypsines.

Discussion

L'examen de nos résultats montre que les paramètres étudiés varient différemment dans les cas de pancréatite, d'infarctus et de coagulopathie de consommation:

1. Dans les infarctus, identiques aux témoins (résultats statistiquement non significatifs);

2. dans les pancréatites, plasminogène légèrement faible, antiplasmines et antitrypsines élevées (résultats significatifs);

3. dans les coagulopathies de consommation, plasminogène abaissé, antiplasmines faibles, antitrypsines élevées (nombre de malades dans chaque cas encore insuffisant pour une analyse statistique).

Certains cliniciens, dont le nombre va croissant, supposent que le pancréas joue un rôle dans beaucoup de syndromes thrombo-emboliques. Nos résultats n'apportent ni confirmation ni infirmation à cette hypothèse. Nous avons antérieurement montré que la pancréatite, dans les premières heures, déclenche une coagulopathie de consommation qui peut, en l'absence de réaction fibrinolytique, se transformer en syndrome thrombo-embolique. Cette CIVD n'est que transitoire.

L'incertitude de la spécificité de la réaction immunologique ne nous permet pas d'affirmer une présence d'enzymes pancréatiques dans le sang circulant au cours d'infarctus ou de CIVD.

Il n'en reste pas moins que la réaction immunologique a été positive dans 92 cas de pancréatite sur 92 et que les taux des inhibiteurs d'antiplasmines et d'antitrypsines ont toujours été trouvés très élevés. Le dosage de ces derniers étant très rapide, il peut être un élément supplémentaire de diagnostic, mais ne doit pas être considéré seul.

Summary

During hypercoagulation syndromes (pancreatitis, infarctus, intravascular disseminated coagulation) the level variations of plasminogen, α_1-antitrypsin, α_2-antiplasmin and the immunodiffusion against an antitrypsin serum are studied. The differences in the results appear to be related to the origin of the hypercoagulation syndrome. Infarctus: Non change; Pancreatitis: Plasminogen slightly low, antiplasmin and antitrypsin significantly high; I.V.D.C.: Plasminogen and antiplasmin low and antitrypsin high.

These results may be helpful for diagnostic purposes.

Adresse des auteurs: Dr. E. G. Vairel et Dr. P. Forlot, Laboratoire Choay, 48, av. Théophile-Gautier, *75 Paris 16*e (France)

7th int. Congr. clin. Chem., Geneva/Evian 1969; vol. 2: Clinical Enzymology, pp. 46–52
(Karger, Basel/München/Paris/New York 1970)

Investigation of Plasma Antithrombin Activity Using ^{131}I-Labelled Human Thrombin

F. Josso, D. Benamon-Djiane, J. M. Lavergne, C. Weilland,
M. Steinbuch
Centre National de Transfusion Sanguine, Paris

We have previously demonstrated that α-2 macroglobulin (α-2 M) is apparently the main carrier of the plasma antithrombin activity [6, 7]. This protein inhibits the clotting activity of thrombin but does not affect its esterolytic activity; it is devoid of heparin-cofactor activity.

Some experiments have suggested that α-2 M may form a complex with thrombin [8]. We decided to investigate the mechanism of the inhibition of thrombin in plasma by using ^{131}I-labelled human thrombin.

Material

Purified human thrombin was obtained by activation of purified prothrombin with a subsequent chromatography on DEAE-Sephadex [1]. This material is homogenous by disc-gel electrophoresis and gives a single precipitation line by immunoelectrophoresis using anti-human plasma antiserum as well as anti-human prothrombin specific antiserum. The thrombin preparation was labelled with radio-iodine according to the chloramine-T method described by Hunter and Greenwood. The material we used had a biological potency of about 1,000 NIH units/mg proteins and a specific activity of 200 μc/mg proteins.

Purified human α-2 macroglobulin was obtained by a method previously described [9]. This material has a potent progressive antithrombin activity.

Semi-purified α-1 antitrypsin (α-1-AT) was obtained by a method combining the following steps: exclusion (pH 5) and adsorption (pH 6) chromatography on DEAE cellulose, precipitation of the α-1–AT containing fraction by 35% ethanol (pH 5; —10°C); gel filtration on Sephadex G100. This material, although completely devoid of α-2 M, has a progressive antithrombin activity but no heparin-cofactor activity.

Rabbit antisera against human prothrombin [3], human α-2 M and total human plasma was obtained in the laboratory. We also used commercial anti-α-1-AT antiserum (Behringwerke).

Experiments

1. Labelled thrombin was added to various media of human origin: defibrinated plasma, albumin, purified α-2 M. The mixtures were submitted to electrophoresis (cellulose acetate) and immunoelectrophoresis (agarose) against various antisera. Then the electrophoresis supports were submitted to autoradiography.

2. Labelled thrombin was inactivated by 0,01 *M* DFP (Diisopropylfluorophosphate) at pH 8,4. In some cases, DFP was first added to thrombin; after 20 min incubation at 37°C, the mixture was adjusted to pH 7,4 and an aliquot then mixed with plasma or α-2 M. In other cases, DFP was added to the incubated mixture thrombin-plasma or thrombin-α-2 M which was again incubated for 20 min and then adjusted to pH 7,4. These various mixtures were submitted to electrophoresis and immunoelectrophoresis with subsequent auto-radiography.

3. α-2 macroglobulin and α-1-AT preparations were mixed after labelled thrombin had been previously added either to the former or to the latter. Then the mixtures were submitted to electrophoresis with subsequent auto-radiography.

Results

1. Binding of thrombin to α-2 M. Thrombin alone migrates as a β-globulin and gives a precipitation line in the same position with antiprothrombin antiserum (fig. 1). In plasma, the radioactivity shows a more anodic (α-2) electrophoretic mobility whereas the thrombin mobility remains unchanged in the presence of purified albumin. When labelled thrombin is added to purified α-2 M the radioactivity migrates with this protein to the same position as in plasma (fig. 2). In plasma as well as in an α-2 M preparation radioactivity bound to thrombin is precipitated by anti-α-2 M antiserum and no longer by anti-prothrombin antiserum (fig. 3).

2. Action of DFP on the binding of thrombin to α-2 M. When thrombin is inactivated by DFP prior to its addition to α-2 M, autoradiography reveals that thrombin retains its original mobility and that it is well precipitated by antiprothrombin antibody. On the contrary, thrombin remains bound to α-2 M if DFP is added after the interaction between thrombin and α-2 M (fig. 4).

3. Apparent binding of thrombin to α-1-AT. When the mixture of labelled thrombin + plasma is submitted to immunoelectrophoresis against anti-human plasma antiserum, the autoradiography reveals the radioactive line of α-2 M and a more anodic spur (α-1) (fig. 5).

When thrombin is first inactivated by DFP, the radioactive line of α-2 M completely disappears but the radioactive spur remains whereas the major part of the radioactivity is revealed in the thrombin position (fig. 5). If

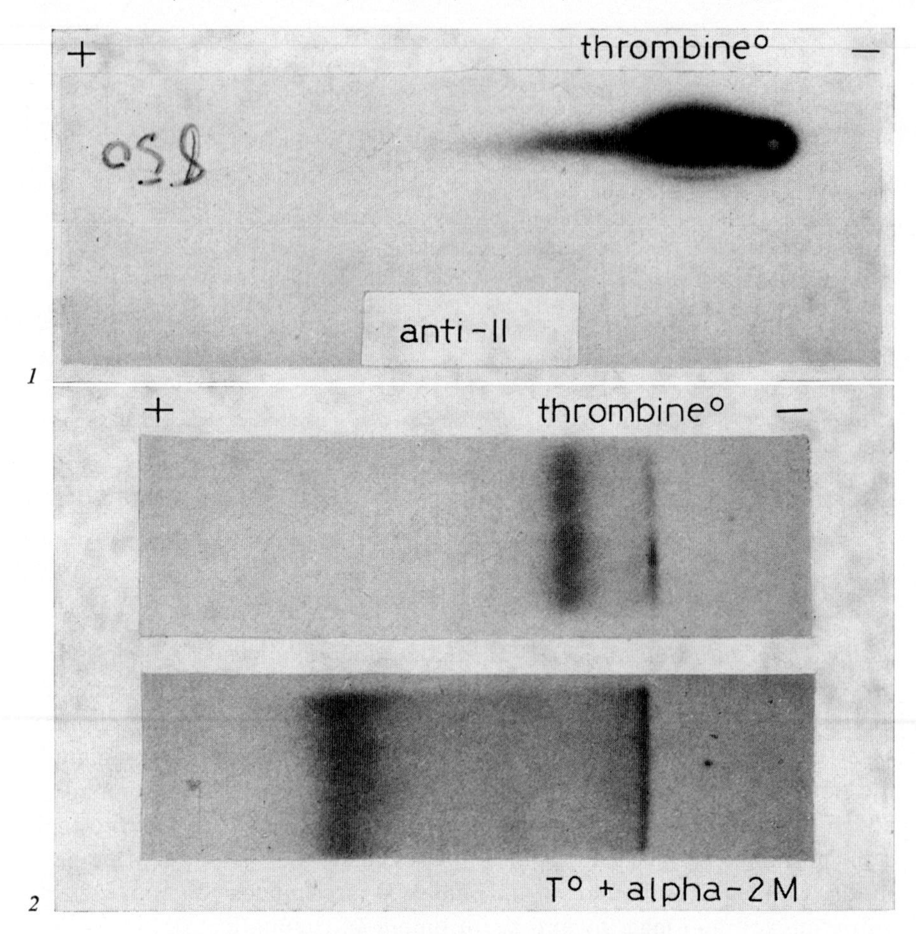

Fig. 1. Immunoelectrophoresis of labelled human thrombin against anti-prothrombin (anti II) anti serum (autoradiography).

Fig. 2. Electrophoresis (cellulose acetate) of labelled thrombin and of a mixture of labelled thrombin (T^0)+α-2 M (autoradiography).

anti-α-1-AT antiserum is used for immunoelectrophoresis of the labelled thrombin-plasma mixture the same anodic spur is revealed alone (fig. 6).

When labelled thrombin is added to an α-1-AT preparation, electrophoresis shows that radioactivity is bound to the α-1-AT band. If α-2 M is added to the mixture, radioactivity is supported by the respective bands of α-1-AT and α-2 M. But if thrombin is added to α-2 M prior to the α-1-AT preparation, all the radioactivity is bound to the α-2 M band.

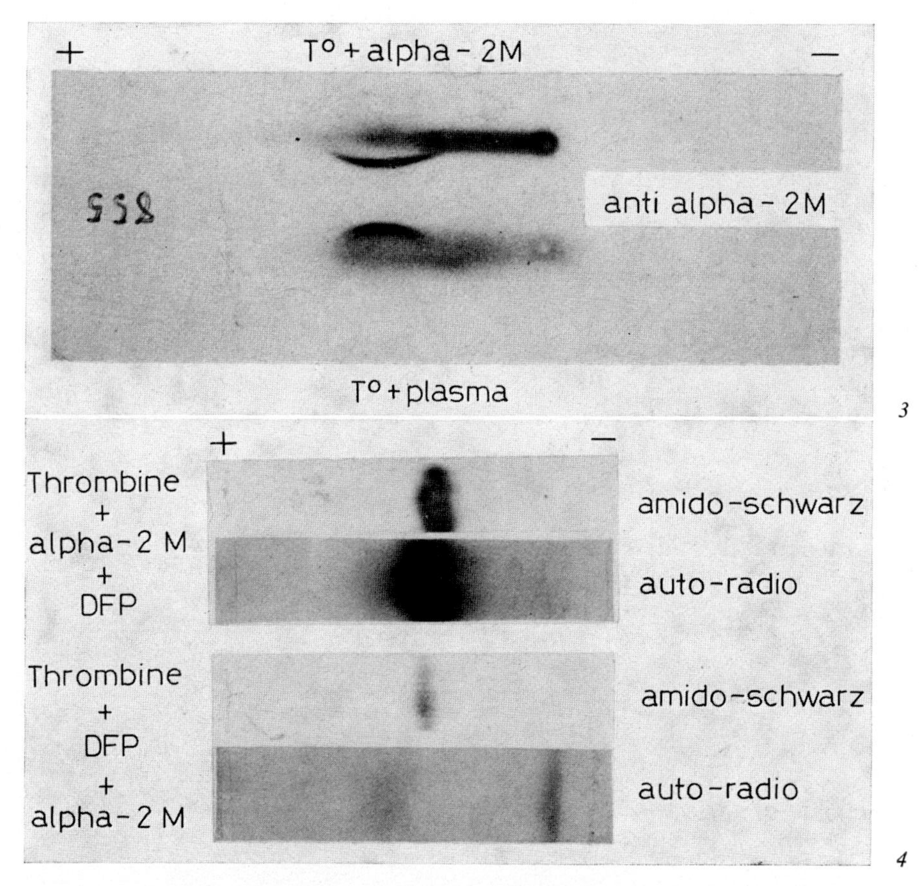

Fig. 3. Immunoelectrophoresis of mixtures of labelled thrombin (T⁰)+α-2 M and +defibrinated plasma against anti α-2 M antiserum (autoradiography).

Fig. 4. Electrophoresis (cellulose acetate) of mixtures of labelled thrombin+DFP+α-2 M. Above, DFP is added after incubation of thrombin with α-2 M; below, α-2 M is added after incubation of thrombin with DFP (Protein coloration and autoradiography).

Discussion

From these experiments it appears that thrombin forms a molecular complex with α-2 M. This finding is in good agreement with our previous observations [7] and those of SHAPIRO and COOPER [5]. In this complex, thrombin keeps its esterolytic activity (i.e. its enzymatic site remains free) but is unable to clot

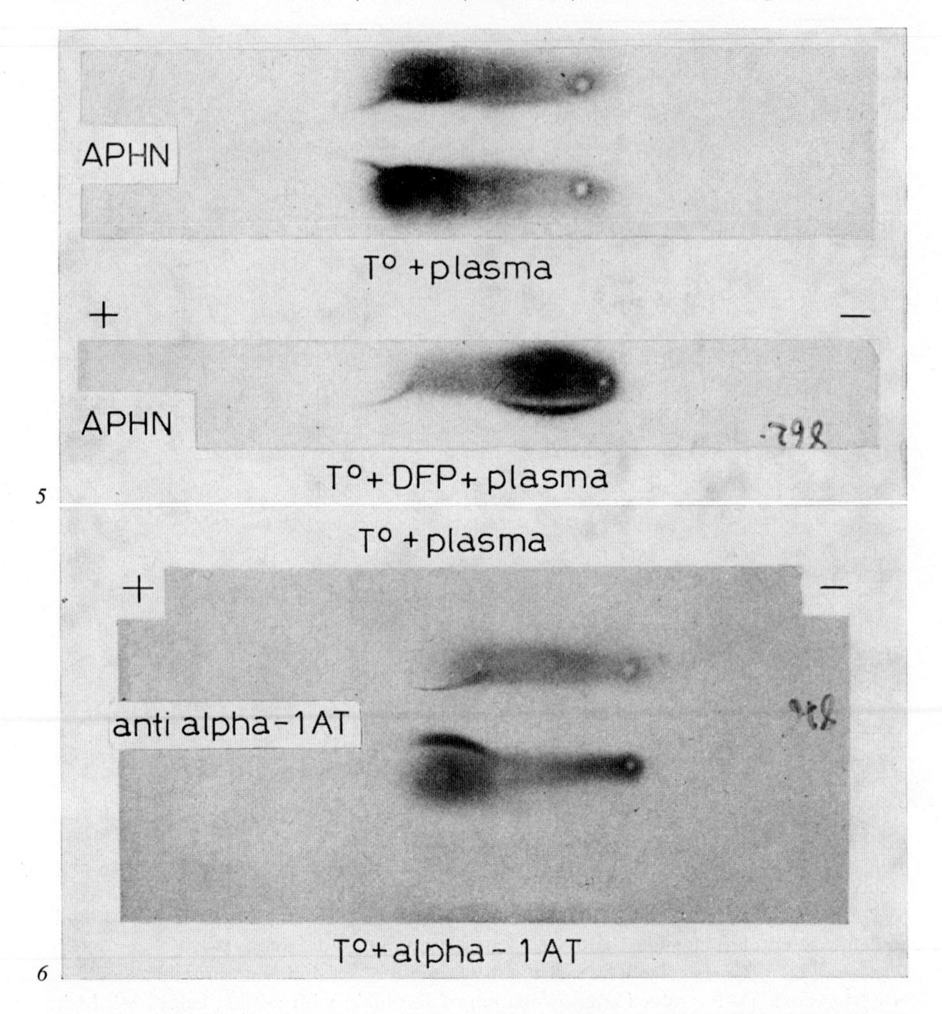

Fig. 5. Immunoelectrophoresis of a mixture of labelled thrombin (T^0)+defibrinated plasma against antiserum to human plasma. Below, thrombin was first inactivated by DFP (autoradiography).

Fig. 6. Immunoelectrophoresis of a mixture of labelled thrombin (T^0)+defibrinated plasma against anti α-1-AT-antiserum (autoradiography).

fibrinogen, undoubtedly by a steric inhibition phenomenon. Moreover the antigenic sites of thrombin are masked in the complex since α-2 M bound thrombin does not react with antiprothrombin antibodies.

The reaction between thrombin and α-2 M needs the enzymatic activity of thrombin since the inhibition of the active site of thrombin by DFP prevents the binding of this molecule by α-2 M. Thus it appears that the inhibition of thrombin by α-2 M involves a proteolytic step in which thrombin probably splits α-2 M before fixing to the modified molecule. This concept would indicate that α-2 M acts as a competitive inhibitor of the action of thrombin on fibrinogen.

Beside the heparin cofactor, α-2 macroglobulin is not the only carrier of plasma progressive antithrombin activity [2]. As it was already postulated [4], it seems that α-1-AT is also able to form a complex with thrombin. The mechanism of the reaction is different from that of the thrombin α-2 M interaction: it is not inhibited by DFP and suppresses the esterolytic activity of thrombin. Under certain experimental conditions, a competition occurs between α-2 M and α-1-AT towards thrombin.

Summary

The mechanism of thrombin inhibition in plasma is investigated by use of radio-iodine labelled human thrombin. In plasma most of the thrombin molecules are bound to α-2 macroglobulin. In this complex, thrombin loses its property to clot fibrinogen but retains its esterolytic activity; the antigenic sites of thrombin are also masked. The formation of the thrombin α-2 macroglobulin complex is inhibited by previous inactivation of the enzymatic site of thrombin by DFP. This finding indicates that α-2 macroglobulin is a substrate of thrombin. A part of thrombin is also bound to the α-1 antitrypsin molecule. The mechanism of this reaction differs from that of the formation of the thrombin α-2 macroglobulin complex. A competition between these two inhibitors towards thrombin may occur.

References

1. BENAMON-DJIANE D., JOSSO F. et SOULIER J. P.: Purification de la prothrombine humaine par chromatographie sur DEAE Sephadex. Propriétés de la prothrombine purifiée. Coagulation *1:* 259 (1969).
2. GANROT P. O.: Electrophoretic separation of two thrombin inhibitors in plasma and serum. Scand. J. clin. Lab. Invest., *24:* 11 (1969).
3. JOSSO F.; LAVERGNE J. M.; WEILLAND C. et SOULIER J. P.: Etude immunologique de la prothrombine et de la thrombine humaines. Thromb. Diath. Haemorrh. *18:* 311 (1967).
4. RIMON A.; SHAMASH Y., and SHAPIRO B.: The plasmin inhibitor of human plasma. IV. Its action on plasmin, trypsin, chymotrypsin and thrombin. J. biol. Chem. *241:*5,102 (1966).

5. SHAPIRO, S. S. and COOPER, J.: Biological activation of radioactive prothrombin. Fed. Proc. *27:* 628 (1968).
6. STEINBUCH M., BLATRIX C. and JOSSO F.: α-2 macroglobulin as progressive antithrombin. Nature *216:* 500 (1967).
7. STEINBUCH M.; BLATRIX C. et JOSSO F.: Action anti-protéase de l'α2-macroglobuline. II. Son rôle d'antithrombine progressive. Rev. franç. Et. clin. biol. *13:* 179 (1968).
8. STEINBUCH, M.; BLATRIX, C.; REUGE C., and JOSSO F.: The mechanism of inactivation of thrombin by α_2 macroglobulin. Proc. XIIth Congr. Intern. Soc. Hemat., New York, 1968 (abstracts, p. 184).
9. STEINBUCH, M.; QUENTIN, M., and PEJAUDIER L.: Specific technique for the isolation of human α_2 macroglobulin. Nature *205:* 1,227 (1965).

Author's address: Dr. F. Josso, Centre National de Transfusion Sanguine, 6, rue Alexandre Cabanel, *Paris 15*e (France)

7th int. Congr. clin. Chem., Geneva/Evian 1969; vol. 2: Clinical Enzymology, pp. 53–56
(Karger, Basel/München/Paris/New York 1970)

Proteaseinhibitoren: Nachweis, Isolierung, Hemmechanismen

H. FRITZ
Institut für Klinische Chemie und Klinische Biochemie der Universität München, München

Proteaseinhibitoren kommen im Tier- und Pflanzenreich überraschend häufig und zum Teil in sehr hoher Konzentration vor [1]. Sie lassen sich in 3 Gruppen einordnen:

1. Die Inhibitoren des Serums mit Molekulargewichten über etwa 50 000. Die wichtigsten von ihnen, α_1-Antitrypsin, α_2-Makroglobulin und Antithrombin III sind imstande, mehrere Proteasen mit ähnlicher Spaltungsspezifität zu inhibieren [2].

2. Die Inhibitoren aus pflanzlichem Material [1, 3, 4] und Eiklar von Vögeln [1, 3] mit Molekulargewichten bis zu etwa 50 000 zeichnen sich durch ihre große Mannigfaltigkeit in ihren Hemmeigenschaften und ihrem Aufbau aus.

3. Die sogenannten Polypeptid-Inhibitoren mit Molekulargewichten bis zu etwa 12 000 sind besonders im Tierreich vertreten. Zu erwähnen sind die Trypsininhibitoren des Pankreas [1, 5], die Trypsin-Chymotrypsin-Plasmin-Inhibitoren der Gl.submand.-Drüsen von Hund und Katze [1, 6], die Trypsin- und Trypsin-Plasmin-Inhibitoren in Samenblasen und Sperma [1, 7], die Trypsin-Plasmin-Inhibitoren und der Thrombininhibitor in Blutegelorganen [8] und der polyvalente Proteaseinhibitor (hemmt u.a. Trypsin, Chymotrypsin, Plasmin und Kallikreine) aus Rinderorganen [1].

Der *quantitative* Nachweis, d.i. die Bestimmung des Inhibitorspiegels in Körperflüssigkeiten und Gewebsextrakten ist immer dann problematisch, wenn sehr niedrige Inhibitorkonzentrationen vorliegen. So kommen z.B. im Pankreassekret auf 100 leicht aktivierbare Trypsinmoleküle nur 1 – 5 Inhibitormoleküle [5]. In diesen Fällen sollten nur solche Hemmtests angewandt werden, bei denen die Kinetik der enzymatischen Reaktion direkt verfolgt werden kann. Die möglichen Fehlerquellen bei Inhibitorbestimmungen in biologischen Flüssigkeiten werden von uns an anderer Stelle ausführlich diskutiert [9].

Die *Isolierung* der Polypeptid-Inhibitoren und der Inhibitoren aus Pflanzensamen und Eiklar von Vögeln aus neutralen bis schwach basischen Salz-Pufferlösungen gelingt auf relativ einfache Weise durch selektive Bindung der Inhibitoren an wasserunlösliche Enzymharze [4, 10]. Nach dem Auswaschen der Begleitsubstanzen lassen sich säurestabile Inhibitoren mit sauren Salz-Pufferlösungen und säurelabile Inhibitoren mit neutralen bis schwach sauren Harnstoff- oder Guanidinsalzlösungen vom unlöslichen Enzymharz-Inhibitor-Komplex wieder ablösen und eluieren. Die Reinheitsgrade der Inhibitoren betragen nach diesem Anreicherungsschritt 60-90% des theoretischen Wertes; durch anschließende Gelfiltration bzw. Ionenaustauschchromatographie erhält man analysenreine Präparate.

Enzymharze mit polyamphoterer Harzstruktur besitzen gegenüber den polyanionischen Enzymharzen günstigere Eigenschaften für die Isolierung: sie besitzen eine höhere Bindungskapazität gegenüber Inhibitoren mit niedrigen isoelektrischen Punkten und die Dissoziation der unlöslichen Komplexe erfolgt bei höheren pH-Werten [11]. Polyanionische und vor allem polyamphotere Inhibitorharze eignen sich auch zur Isolierung von Plasmin und von Kallikreinen [12].

Die Inhibitoren unterscheiden sich nicht nur in der Zahl der Enzyme, die sie hemmen ('Hemmspektrum'), sondern auch im *Inhibierungsmechanismus*. So kennen wir z.Zt. 3 verschiedene Hemmtypen:

1. Die *permanente Hemmung* mit folgenden Kriterien:
   a) Das Hemmgleichgewicht ist in Sekunden bis wenigen Minuten erreicht (Sofort-Hemmung),
   b) die Geschwindigkeit der Gleichgewichtseinstellung ist relativ temperaturunabhängig,
   c) der erreichte Hemmgrad bleibt über Stunden und Tage unverändert,
   d) die Bildung des Enzym-Inhibitor-Komplexes ist reversibel.

Permanente Inhibierung verursachen z.B. nahezu alle Inhibitoren aus pflanzlichem Material, der polyvalente Proteaseinhibitor aus Rinderorganen, und von den Serum-Inhibitoren α_1-Antitrypsin gegenüber Trypsin und α_2-Makroglobulin gegenüber Plasmin.

2. Bei der *temporären Hemmung* nimmt im Gegensatz zur permanenten Hemmung der Hemmgrad (Kriterium c; a, b und d wie oben) bei der Inkubation des Komplexes mit der Zeit ab, d.h. das inhibierte Enzym wird infolge eines enzymatischen Abbaus des Inhibitors langsam freigesetzt. Temporär hemmen die meisten der Polypeptid-Inhibitoren, so die Inhibi-

toren aus Pankreas, Samenblasen und Sperma, Gl. submand. und die Trypsin-Plasmin-Inhibitoren aus Blutegelorganen.

3. Für die *Progressiv-Hemmung* gelten folgende Kriterien:
   a) Das Hemmgleichgewicht wird erst nach mehreren Stunden bei einer Inkubationstemperatur von 25–37° C erreicht,
   b) die Geschwindigkeit der Gleichgewichtseinstellung ist stark von der Inkubationstemperatur abhängig,
   c) die Bildung des 'Komplexes' ist irreversibel.

Progressives Hemmverhalten ist bisher nur bei Inhibitoren des Serums gefunden worden, und zwar werden progressiv gehemmt: Plasmin durch α_1-Antitrypsin [13], Thrombin durch Antithrombin III [13, 14] und Kallikrein aus Schweineserum durch den Trypsin-Plasmin-Inhibitor aus Rinderserum [15]. Nach unseren Untersuchungen ist das α_1-Antitrypsin aus Humanserum mit dem Progressiv-anti-Kallikrein identisch [16]:1 ml Humanserum inaktiviert demnach in 5 Stunden bei 37° C 1500 biologische Kallikreineinheiten.

Summary

Besides the proteinase inhibitors in animal sera and plant tissues with higher molecular weights polypeptid inhibitors (Mol. W. 3,000–12,000) are widly distributed in animal tissues; e.g. the trypsin inhibitors of pancreas glands, the trypsin-chymotrypsin-plasmin inhibitors in gl. submand. of dogs, the trypsin and trypsin-plasmin inhibitors in seminal vesicles and the trypsin-plasmin inhibitors and thrombin inhibitor in leeches. The estimation of these inhibitors in extracts and body fluids, their isolation with water insoluble enzyme resins and possible inhibition mechanisms are discussed. α_1-Antitrypsin of human serum is identical with the progressiv inhibitor of Kallikrein.

Literatur

1. VOGEL, R.; TRAUTSCHOLD, I. und WERLE, E.: Natürliche Proteinasen-Inhibitoren (Thieme, Stuttgart 1966); VOGEL, R.; TRAUTSCHOLD, I. und WERLE, E.: Natural Proteinase Inhibitors (Academic Press, New York/London, 1968).
2. HEIMBURGER, N.; und HAUPT, H.: Klin. Wschr. *44:* 1 196–1 199 (1966).
3. FEENY, R. E. und ALLISON, R. G.: Evolutionary Biochemistry of Proteins (Wiley-Interscience, New York, 1969), S. 199–244.
4. HOCHSTRASSER, K.; WERLE, E.; SIEGELMANN, R. und SCHWARZ, S.: Z. physiol. Chem. *350:* 897 (1969); HOCHSTRASSER, K; ILLCHMANN, K. und WERLE, E.: Z. physiol. Chem. *350:* 655 (1969); HOCHSTRASSER, K. und WERLE, E.: Z. physiol. Chem. *350:* 249 (1969); HOCHSTRASSER, K.; SCHWARZ, S.; ILLCHMANN, K. und WERLE, E.:

Z. physiol. Chem. *349:* 1 449 (1968); HOCHSTRASSER, K.; MUSS, M. und WERLE, E.: Z. physiol. Chem. *348:* 1 337 (1967).

5. FRITZ, H.; HÜLLER, I.; WIEDEMANN, M. und WERLE, E.: Z. physiol. Chem. *348:* 405 (1967); FRITZ, H.; TRAUTSCHOLD, I.; HAENDLE, H. und WERLE, E.: Ann N.Y. Acad. Sci. *146:* 400 (1968); FRITZ, H.; HUTZEL, M.; HÜLLER, I.; WIEDEMANN, M.; STAHLHEBER, H.; LEHNERT, P. und FORELL, M.-M.: Z. physiol. Chem. *348:* 1 575 (1967); FRITZ, H.; HOCHSTRASSER, K. und WERLE, E.: Z. analyt. Chem. *243:* 452 (1968); HOCHSTRASSER, K.; SCHRAMM, W.; FRITZ, H.; SCHWARZ, S. und WERLE E.: Z. physiol. Chem. *350:* 893 (1969); GREENE, L. J.; DICARLO, J. J.; SUSSMAN, A. J. und BARTELT, D. C.: J. biol. Chem. *243:* 1 804 (1968); GREENE, L. J.: Ann. N.Y. Acad. Science *146:* 387 (1968); GREENE, L. J.; RIGBI, M. und FACKRE, D. S.: J. biol. chem. *241:* 5 610 (1966); GREENE, L. J.; GIORDANO, J. S.: J. biol. Chem. *244:* 285 (1969); GREENE, L. J.; BARTELT, D. C.: J. biol. Chem. *244:* 2 646 (1969); CERWINSKY, E. W.; BURCK, P. J. und GRINNAN, E. L.: Biochemistry *6:* 3 175 (1967).
6. TRAUTSCHOLD, I.; WERLE, E.; HAENDLE, H. und SEBENING, H.: Z. physiol. Chem. *332:* 328 (1963); FRITZ, H.; PASQUAY, P.; MEISTER, R. und WERLE, E.: Z. physiol. Chem. (im Druck).
7. HAENDLE, H.; FRITZ, H.; TRAUTSCHOLD, I. und WERLE, E.: Z. physiol. Chem. *343:* 185 (1965).
8. FRITZ, H.; OPPITZ, K.-H.; GEBHARDT, M.; OPPITZ, I. und WERLE, E.: Z. physiol. Chem. *350:* 91 (1969); FRITZ, H.; GEBHARDT, M.; FINK, E. und WERLE, E.: Z. physiol. Chem. (im Druck).
9. FRITZ, H.; TRAUTSCHOLD, I. und WERLE, E.: in BERGMEYER, Methoden der Enzymatischen Analyse (Verlag Chemie, 1970).
10. FRITZ, H.; SCHULT, H.; NEUDECKER, M. und WERLE, E.: Angew. Chem. *78:* 775 (1966); FRITZ, H.; SCHULT, H.; HUTZEL, M.; WIEDEMANN, M. und WERLE, E.: Z. physiol, Chem. *348:* 308 (1967).
11. FRITZ, H.; GEBHARDT, M.; FINK, E.; SCHRAMM, W. und WERLE, E.: Z. physiol. Chem. *350:* 129 (1969).
12. FRITZ, H.; BREY, B.; SCHMAL, A. und WERLE, E.: Z. physiol. Chem. *350:* 617 (1969).
13. HEIMBURGER, N. und HAUPT, H.: Klin. Wschr. *44:* 1 196 (1966).
14. HEIMBURGER, N.: First International Symposium on Tissue Factors in Homeostasis of the Coagulation-Fibrinolysis System, Firenze, Maggio 1967, S. 353.
15. HABERMANN, E.: Ann. N.Y. Acad. Sci. *146:* 479 (1968).
16. FRITZ, H.; BREY, B. und SCHMAL, A.: Z. physiol. Chem. *350:* 1551 (1969).

Adresse des Autors: Dr. H. FRITZ, Institut für Klinische Chemie und Klinische Biochemie der Universität München, *D-8 München* (Deutschland)

7th int. Congr. clin. Chem., Geneva/Evian 1969; vol. 2: Clinical Enzymology, pp. 57–60
(Karger, Basel/München/Paris/New York 1970)

Hemmzentren von Proteaseninhibitoren

K. HOCHSTRASSER
Institut für Klinische Chemie und Klinische Biochemie der Universität München, München
(Direktor: Prof. Dr. Dr. E. WERLE)

Die Arbeitsgruppen um ACHER [1, 2] und LASKOWSKY sen. [2, 4] konnten am Beispiel des polyvalenten Proteaseninhibitors aus Rinderlunge zum ersten Mal nachweisen, daß die Hemmwirkung dieses Inhibitors offensichtlich durch einen Lysinrest an definierter Stelle der Polypeptidkette gewährleistet wird. Die Umsetzung einmal des freien Inhibitors und des im Trypsininhibitorkomplex gebundenen Hemmstoffs mit dem Carboxyanhydrid des Alanins führt im ersten Fall zur Inaktivierung des Inhibitors, während im zweiten Fall nach Abtrennung des Trypsins aus dem Komplex aktiver Inhibitor zurückerhalten wird. Die Untersuchung beider Produkte ergab, daß im Enzyminhibitorkomplex der Lysinrest in Stellung 15 nicht acyliert wird, also gegen den Angriff des Acylierungsmittels geschützt ist. Diese Tatsache ist nur zu erklären, wenn angenommen wird, daß die Wechselwirkung des Inhibitors mit dem Enzymmolekül von diesem Lysinrest ausgeht. Dieser Lysinrest wird deshalb folgerichtig als aktives Zentrum des Inhibitors angesehen.

Die Bedeutung dieses Lysinrestes für die Hemmwirkung konnte LASKOWSKY sen. auf völlig anderem Weg nachweisen. Er führte eine partielle Reduktion einer von einem dem Lysinrest benachbarten Cysteinrest ausgehenden S-S-Brücke durch und fixierte die Thiolgruppe durch Carboxymethylierung einmal mit Jodessigsäure und einmal mit Jodacetamid. Das Produkt aus der Umsetzung mit Jodacetamid ist noch aktiv, während das Carboxymethylierungsprodukt aus Jodessigsäure keine Hemmwirkung besitzt. Es wird angenommen, daß sich zwischen der neu eingeführten Carboxylgruppe und der ε-Aminogruppe des Lysins eine Ionenbindung ausbildet, die die ε-Aminogruppe des Lysins, das als aktives Zentrum angesehen wird, blockiert.

Von LASKOWSKY jr. [5, 6] durchgeführte Untersuchungen ergaben, daß Trypsin bei niederen H-Ionenkonzentrationen in der Lage ist, mit besonderen reaktiven Peptidbindungen der Trypsininhibitoren aus Sojabohnen und Ovomucoid unter Spaltung einer Arg-Ileu- bzw. Arg-Ala-Bindung zu reagieren. Die aus der Reaktion hervorgehenden Produkte werden als modifizierte Hemmstoffe bezeichnet, sie sind voll aktiv, verlieren jedoch ihre Hemmwirkung sofort, wenn durch Carboxypeptidase B der an der Peptidbindung beteiligte Argininrest abgespalten wird. In diesen Fällen werden die durch Trypsineinwirkung freilegbaren Argininreste als aktive Zentren dieser Inhibitoren deklariert.

Durch das von FRITZ und WERLE [7] entwickelte Verfahren der reversiblen Bindung von Trypsininhibitoren an wasserunlösliche Enzymharze war es uns möglich, eine Reihe von Inhibitoren aus tierischem und pflanzlichem Material erstmals relativ rasch und einfach in hoher Reinheit zu gewinnen [8, 9, 10, 11, 12, 13]. Die nähere Untersuchung der auf diesem Weg isolierten Hemmstoffe ergab überraschend, daß die meisten Inhibitoren durch das angewandte Verfahren, das ja als wesentlichen Reinigungsschritt eine Komplexierung mit aktivem Trypsin beinhaltet, in modifizierter Form erhalten werden Diese Inhibitoren besitzen alle 2 N-terminale Aminosäurereste, sie bestehen aus 2 durch S-S-Brücken verbundenen Peptidketten. Nach Reduktion und Fixierung der SH-Gruppen ist es möglich, die modifizierten Inhibitoren in inhibitorisch unwirksame Ketten zu zerlegen. Jeweils eine der Ketten enthält carboxylendständiges Arginin oder Lysin, offensichtlich entstanden durch Lösen einer reaktiven Peptidbindung bei der Komplexierungsreaktion. Abspaltung der carboxylendständigen basischen Aminosäurereste der modifizierten Inhibitoren führt zum Verlust der Hemmwirkung. Die durch dieses Verfahren aufgefundenen reaktiven Peptidbindungen der Inhibitoren stellen offensichtlich somit auch zugleich das sogenannte aktive Zentrum dar. Durch Bestimmung der Längen der einzelnen Peptidketten ist die einfache Lokalisierung dieser reaktiven Zentren möglich, außerdem erlaubt die direkte Bestimmung der C-terminalen Aminosäure eine sofortige Klassifizierung der Hemmstoffe als sogenannte Lysin- oder Arginin-Inhibitoren.

Einen weiteren direkten Beweis für unsere Ansicht, daß diese reaktiven Zentren auch das aktive Zentrum des jeweiligen Inhibitors darstellen, sehen wir darin, daß es möglich ist, die bei diesem Vorgehen als Lysin-Inhibitoren erkannten Hemmkörper reversibel durch Acylierung zu inaktivieren [14]. Arginin-Inhibitoren werden in ihrem Hemmverhalten durch Acylierung der α- und ε-Aminogruppen nicht beeinflußt, sie sind jedoch inaktivierbar durch gruppenspezifische Umsetzungen an den Guanidogruppen der Argininreste [14], z.B. mittels Butandion. LASKOWSKY lieferte für diese Auffassung über das aktive Zentrum der Trypsininhibitoren einen eindrucksvollen Beweis, indem es ihm gelang, den Arginininhibitor aus Sojabohnen nach enzymatischer Abspaltung des Argininrestes auf enzymatischem Weg in einen Lysininhibitor [15] umzuwandeln.

Die Zusammenfassung der bisher vorliegenden Ergebnisse erlaubt folgende Aussagen zu machen. Die Wirkung der als Trypsinhemmstoffe klassifizierten Proteine beruht darauf, daß diese Eiweißkörper definierte reaktive Peptidbindungen enthalten, die als Substrate für das aktive Zentrum des Trypsins zugänglich sind, alle übrigen an sich von Trypsin spaltbaren Peptidbindungen sind infolge besonderer sterischer Verhältnisse im Inhibitormolekül geschützt. Die Hemmwirkung kann als Proteolyse interpretiert werden, die auf verschiedenen Stufen des Proteolysevorganges unterbrochen sein kann. Beim Beispiel der Inhibitoren aus Rinder- oder Schaflunge, die ohne Modifizierung aus den Komplexen zurückerhalten werden, kann angenommen werden, daß diese limitierte Proteolyse über die Bildung eines Michaeliskomplexes nicht hinausgeht. Die Bildung von Acylenzymkomplexen scheint bei verschiedenen Pankreasinhibitoren mit nur teilweiser Modifizierbarkeit vorzuliegen, während bei den vollständig modifizierbaren und in dieser Form voll aktiven pflanzlichen Inhibitoren Produktenzymkomplexe gebildet werden dürften.

Die Funktionsfähigkeit der sogenannten reaktiven Peptidbindungen als Hemmzentrum ist nach unseren Untersuchungen in der besonderen Raumstruktur [16] der Hemmstoffe zu sehen. Wie wir zeigen konnten, geht die permanente Hemmwirkung z.B. der pflanzlichen Inhibitoren schon bei geringster Störung der Raumstruktur, die durch S-S-Brücken bedingt ist, verloren. Komplexe dieser Inhibitoren mit Trypsin zerfallen in Gegenwart geringer Thiolmengen bereits in harnstoffreien Lösungen sofort, d.h. es wird aktives Trypsin freigesetzt und der Hemmstoff durch Trypsin angreifbar. Sequenzuntersuchungen von GREEN [17] und uns an den speziellen Trypsininhibitoren aus Rinder-, Schaf [13]- und Schweinepankreas [18], Granineensamen und Meerschweinchensamenblasen ergaben als Gemeinsamkeit, daß in allen diesen Hemmstoffen relativ lange etwa 20 Aminosäurereste umfassende Aminosäuresequenzen vorkommen, die keine Lysin- oder Argininreste enthalten. Die Bedeutung dieser Tatsache wird wohl erst voll zu erkennen sein, wenn die Raumstruktur dieser Verbindungen bekannt ist. Es kann jedoch vorerst angenommen werden, daß die reaktiven, als Hemmzentren fungierenden Peptidbindungen, an der Oberfläche der

Moleküle liegen und die übrigen durch Trypsin angreifbaren Bindungen im Inneren der Moleküle durch Umhüllung mit den angeführten langen gegen Trypsin stabilen Aminosäuresequenzen geschützt sind.

Summary

Trypsin inhibitors of animal and plant tissues contain so-called reactive peptide linkages at definitive positions of the peptide chain. The linkages of the type Arg-X, respectively Lys-X are seen as active centers of these inhibitors. Methods are described, which allow the location of these reactive peptide bounds in the amino acid sequence. Additionally it is explained, that the conformation is important for this inhibitor effect.

Literatur

1. ACHER, R.; CHAUVET, I.; ARNON, R. und SELA, M.: Europ. J. Biochem. *3:* 476 (1968).
2. CHAUVET, I. und ACHER, R.: J. biol. Chem. *242:* 4,274 (1967).
3. KRESS, L.F. und LASKOWSKY, M.: Sen. J. Biol. Chem. *242:* 4,925 (1967).
4. KRESS, L. F.; WILSON, K. A. and LASKOWSKY, M.: Sen. J. biol. Chem. *243:* 1,758 (1968).
5. FINKENSTADT, W. R. and LASKOWSKY, M., Jr.: J. biol. Chem. *240:* PC 962 (1965).
6. FINKENSTADT, W. R. and LASKOWSKY, M., Jr.: J. biol. Chem. *242:* 771 (1967).
7. FRITZ, H.; HÜLLER, I.; HUTZEL, M.; WIEDEMANN, M. and WERLE, E.: Z. physiol. Chem *348:* 405 (1967).
8. HOCHSTRASSER, K.; MUSS, M. and WERLE, E.: Z. physiol. Chem. *348:* 1,337 (1967).
9. HOCHSTRASSER, K.; WERLE, E.: Z. physiol. Chem. *350:* 249 (1969).
10. HOCHSTRASSER, K. and WERLE, E.: Z. physiol. Chem. *350:* 897 (1969).
11. HOCHSTRASSER, K.; ILLCHMANN, K. and WERLE, E.: Z. physiol. Chem. *350:* 929 (1969).
12. HOCHSTRASSER, K. and WERLE, E.: Z. physiol. Chem. *350:* 249 (1969).
13. HOCHSTRASSER, K.; SCHRAMM, W.; FRITZ, H.; SCHWARZ, S. and WERLE, E.: Z. physiol. Chem. *350:* 893 (1969).
14. FRITZ, H.; FINK, E. GEBHARDT, M.; HOCHSTRAßER, K.; WERLE, E.: Z. physiol. Chem. *350:* 933 (1969).
15. SEALOCH, R. B. and LASKOWSKY, M., Jr.: Biochemistry (Worthington), (i. Druck).
16. HOCHSTRASSER, K.; SCHWARZ, S.; ILLCHMANN, K. and WERLE, E.: Z. physiol. Chem. *349:* 1,449 (1968).
17. GREENE, L. I. and GIORDANO, I. S., Jr.: J. biol. Chem. *244:* 285 (1969).
18. TSCHESCHE, H.: Z. physiol. Chem. *350:* (i. Druck).

Adresse des Autors: Dr. K. HOCHSTRASSER, Institut für Klinische Biochemie der Universität München, *D-8 München* (Deutschland)

7th int. Congr. clin. Chem., Geneva/Evian 1969; vol. 2: Clinical Enzymology, pp. 61–66
(Karger, Basel/München/Paris/New York 1970

Variation of the Activity of α_2-macroglobulin as Progressive Antithrombin after Molecular Modification

M. STEINBUCH and C. REUGE
National Center for Blood Transfusion, Paris

α_2-macroglobulin (α_2M) is a potent inhibitor of several proteases such as trypsin, plasmin, thrombin and elastase. Most inhibitors react directly with the active site of these proteolytic enzymes and the complexes formed between these inhibitors and the enzymes show no more activity whether the substrate is a high molecular protein or a small synthetic molecule like TAMe[1] or BAEe[2]. The interaction occurring between α_2M and these enzymes is not of the same type. Indeed, the complexes resulting from the interaction between the protease and α_2M still have esterase activity, whereas their proteolytic activity against high molecular substrates is abolished.

α_2M loses its antiprotease activity [1] by dissociation of the molecule [2] when the frozen solution is kept for several weeks at —15° C or at room temperature in the presence of 0.25M methylamine [3]. The present study deals with the anti-protease activity of chemically modified α_2M. Progressive antithrombin activity has been chosen for the evaluation of the residual activity because of the simplicity and accuracy of this method [4].

Materials and Methods

A. Modifying Reactions

α_2M is prepared from human ACD-plasma by a technique involving adsorption of contaminating plasminogen by bentonite, precipitation of an α_2M rich fraction by

[1] TAMe: Tosyl-arginine-methyl ester.
[2] BAEe: Benzoyl-arginine-ethyl ester.

rivanol and final batch-adsorption of the remaining impurities on DEAE-cellulose at pH 5 and 0.015M acetate concentration. The pure α_2M is then modified by the following reactions:

1. Pure α_2M is incubated at pH 5.7 in acetate buffer containing 0.05% $CaCl_2$ with 10 U neuraminidase (Behringwerke, Marburg/Germany) per mg protein for 30h at 37° C. A control preparation is kept under the same conditions of pH etc. at 37° C. Both samples are finally dialyzed for 48 h.

2. α_2M is oxidized at pH 4.6 in the presence of a) 0.005M $NaIO_4$ and b) 0.01M $NaIO_4$. One series is left for 40 min at 4° C in the dark in contact with the reagent, the other for 3h; all samples are then dialyzed against buffered saline. Control samples are prepared as above.

3. α_2M is substituted at pH 9 with bromo-acetic acid following the technique of KORMAN and CLARKE [5]. Control and dialysis as already mentioned.

4. α_2M is succinylated with succinic anhydride at pH 8 as described by HABEEB *et al.* [6]. The amount of the reagent used being 20–60% of the weight of α_2M to be succinylated. The excess reagent is removed by filtration on sephadex G_{25} followed by dialysis.

5. α_2M is treated at pH 7 with N-bromosuccinimide for 4 h at 4° C. For 100 mg of α_2M various amounts from 4–10 mg NBS are used. The samples are filtered on G_{25} and thoroughly dialyzed.

B. Analytical Procedures

1. Electrophoretic analysis, on cellulose acetate strips and on starch gel (micro-technique).

2. Ultracentrifugation at the 'Station Centrale d'Ultracentrifugation du C.N.R.S.' using a Spinco model E Centrifuge.

3. Evaluation of the sialic acid content by the technique of WARREN [7].

4. Titration of the free NH_2-group with the ninhydrin reagent [8].

5. Titration of the tryptophan residues [9].

6. Determination of the progressive antithrombin activity of the sample using bovine thrombin (Roche) [4.].

Results

1. The removal of 90% sialic acid by sialidase does not modify the progressive antithrombin activity of α_2M. The desialized sample migrates like a β_1-globulin (fig. 1).

2. Oxidation with $NaIO_4$ 0.01 M for 40 min at 4° C results in the loss of about 30% of antithrombin activity although the sample still has more (25%) sialic acid than the sialidase treated sample. When α_2M is oxidized in the presence of 0.005 M $NaIO_4$ 40% of sialic acid is left intact and only about 10% of the activity lost.

3. Carboxymethylation results in the substitution of about half of the available amino groups but only 15% of the activity is lost under these

conditions. Electrophoresis shows a somewhat heterogeneous broad band in the α_1/albumin region (fig. 1).

4. The most spectacular results are observed after extensive succinylation of α_2M. Most of the activity (85%) is lost in samples having 90% of their available amino groups blocked by the substituting reagent (table I). Complete dissociation of the molecule is observed in the ultra-centrifuge. The transformation of the macromolecule into subunits is always observed. However, the appearance of a main constituent (85%) of 5.5-6.5 S is most frequently noticed, whereas the dissociation in an even smaller fragment of 2.1 S is only found occasionally. The substituted protein has a faster mobility ($\approx$40%) than albumin in cellulose acetate strips but is poorly stainable. Furthermore, the succinylated α_2M lost its immunological reactivity.

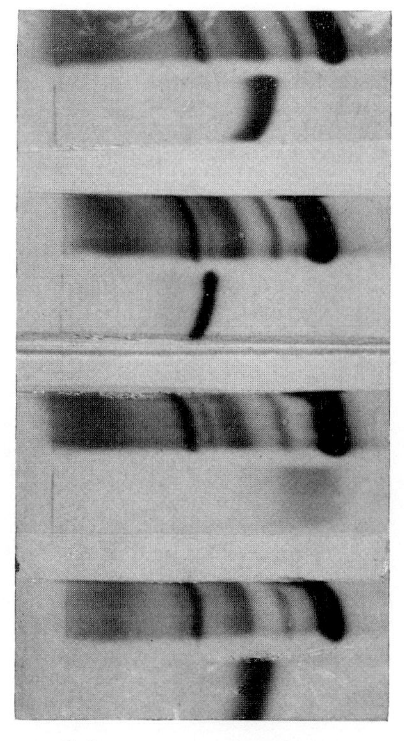

Fig. 1. Electrophoresis on cellulose acetate strips of α_2M and derivatives. 1, 3, 5, 7 = serum control; 2 = α_2M control; 4 = α_2M treated by sialidase; 6 = carboxymethylated α_2M; 8 = α_2M oxidized by NBS.

Table I. Progressive antithrombin activity of succinylated α_2M

Sample	Thrombin units time : 0	Thrombin units time : 30 min
1	25	19
2	25	2
3	25	25

1 = Succinylated sample (1.2%); 90% of the available amino groups are blocked; most of the activity is lost; 2 = control sample of α_2M (1.1%); 3 = thrombin buffer control showing that the loss of activity is not due to the incubation per se.

5. Oxidation of α_2M by NBS gives similar but less striking results: 80% of the activity is lost when 40% of the tryptophan residues are destroyed whereas only 15% of the activity is lost when 25% of the tryptophan residues are oxidized. The electrophoretic mobility of the modified samples is unchanged in cellulose acetate strips (fig. 1) but two well separated bands are observed in starch gel electrophoresis (fig. 2) and the ultracentrifugation pattern revealed that 50% of the preparation is dissociated to give an 11 S component leaving the other half of the preparation in the original macromolecular state.

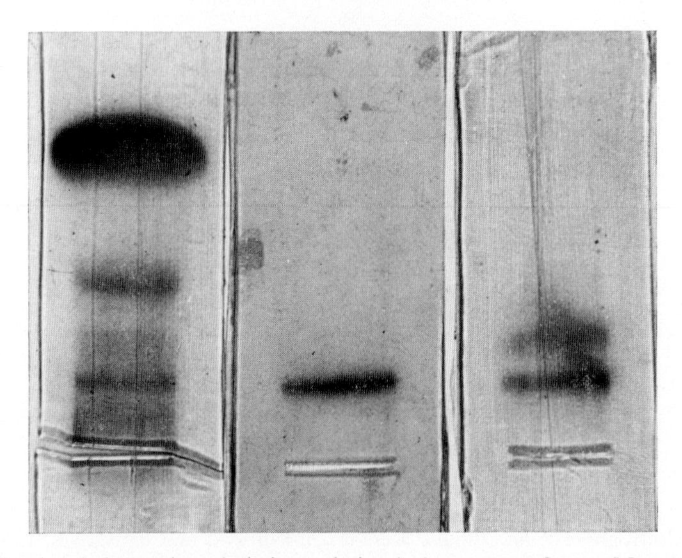

Fig. 2. Starch gel electrophoresis (microtechnique). 1 = serum; 2 = α_2M control; 3 = α_2M oxidized by N-bromosuccinimide.

Discussion

The experiments with sialidase show that sialic acid is not relevant for the activity of α_2M as a progressive antithrombin. Oxidation by $NaIO_4$ is less specific than the enzyme treatment. Other groupings than sialic acid may well be involved and thus explain the partial loss of activity observed in these samples.

Dissociation of the molecule as observed for the succinylated samples and (to a lesser degree) for the α_2M oxidized by NBS indicate that this macroglobulin is set up by subunits. Similar results have been reported for succinylated lipoprotein [10] and for succinylated ceruloplasmin [11] for instance. Thus the integrity of the quaternary structure seems to be essential for the antiprotease activity of α_2M.

Summary

Pure human α_2-macroglobulin was submitted to various chemical modifications. Sialic acid has been removed by incubation with sialidase without modification of the antiprotease activity whereas oxydation with periodate gave a slightly different result. Furthermore the inhibitor was modified by succinylation, carboxymethylation and by oxydation with N – bromosuccinimide. The most striking result was observed after extensive succinylation as the substituted protein was dissociated into subunits while 80% of the antithrombin activity was lost.

References

1. Steinbuch, M. et Blatrix, C.: Action anti-protéase de l'α_2-macroglobuline. I. Activités antiplasmine et antitrypsine. Rev. franc. Et. clin. biol. *13:* 142–152 (1968).
2. Gentou, Cl.: Relation entre les structures tertiaire et quaternaire des macroglobulines humaines normales. C.R.Acad. Sci. *260:* 6468 (1965).
3. Steinbuch, M.; Pejaudier, L.; Quentin, M. and Martin, V.: Molecular alteration of α_2-macroglobulin by aliphatic amines. Biochim. biophys. Acta *154:* 228–231 (1968).
4. Steinbuch, M.; Blatrix, C. et Josso, F.: Action anti-protéase de l'$\alpha_2$macro-globuline. II. Son rôle d'antithrombine progressive. Rev. franç. Et. clin. biol. *13:* 179–186 (1968).
5. Korman, S. and Clarke, H. T.: Carboxymethyl proteins. J. biol. Chem. *221:* 133. (1956).
6. Habeeb, A. F.; Cassidy, H. G. and Singer. S. J.: Molecular and structural effects produced in proteins by reaction with succinic anhydride. Biochim. biophys. Acta. *29:* 587 (1958).
7. Warren, L.: The thiobarbituric acid assay of sialic acids. J. biol. Chem. *234:* 1971 1975 (1969).

8. TROLL, W. and CONNAN, R. K.: A modified photometric ninhydrin method for the analysis of amino acids and imino acids. J. biol. Chem. *200:* 803 (1953).
9. SAIFER, A. and GERSTENFELD, S.: The spectrophotometric determination of the albumin-globulin ratio of biological fluids with a modified tryptophan reaction. Clin. Chem. *7:* 563–564 (1961).
10. SCANU, A.; POLLARD, H. and READER, W.: Properties of human serum low density lipoproteins after modification by succinic anhydride. J. Lipid Res. *9:* 342–349 (1968).
11. POILLON, W. N. and BEARN, G.: The molecular structure of human ceruloplasmin: evidence for subunits. Biochim. biophys. Acta *127:* 407–427 (1966).

Authors' address: Dr. M. STEINBUCH and Dr.C. REUGE, Centre National de Transfusion Sanguine, *Paris 15*e (France)

7th int. Congr. clin. Chem., Geneva/Evian 1969; vol. 2: Clinical Enzymology, pp. 67–73
(Karger, Basel/München/Paris/New York 1970)

Sur la formation d'un complexe entre les immunoglobulines IgA du colostrum humain et la trypsine

Y. Counitchansky, G. Berthillier et R. Got
Laboratoire de Chimie biologique, Faculté de Médecine, Lyon

Si l'on filtre, sur une colonne de Sephadex G 200, un mélange de colostrum total et de trypsine, l'essentiel de l'activité trypsique, dosée par hydrolyse de l'ester méthylique de la *p*-tosyl-L-arginime (TAME) se retrouve dans les protéines exclues du Sephadex [1]. On peut donc en déduire que la trypsine forme, avec certaines protéines de colostrum, des complexes à activité estérasique, de poids moléculaire supérieur à 200 000.

La centrifugation en gradient de saccharose (5–20%), effectuée selon la méthode de Britten et Roberts [2] dans une centrifugeuse Spinco avec le rotor SW 39, confirme ce résultat. En effet, alors que la trypsine libre migre vers le tiers supérieur du tube, une partie de l'activité se retrouve au fond du tube, avec les protéines possédant un poids moléculaire élevé.

La présence de protéines complexantes dans les fractions lourdes du colostrum suggère qu'il s'agit d'une macroglobuline et l'on pense évidemment à l'α_2-macroglobuline du sérum. Bien que l'on trouve effectivement, dans le colostrum humain, toutes les protéines du sérum sanguin [3], la teneur en α_2-macroglobuline semble trop faible pour que l'on puisse la rendre responsable de cette activité. Aussi avons-nous cherché à isoler et caractériser ces protéines actives afin de les identifier.

Les laits colostraux sont délipidés par 30 min de centrifugation à 105 000 g. Les protéines sont précipitées, à —10°C, par un volume d'alcool et le précipité est extrait par un tampon phosphate 0,1 M pH 7.

Cet extrait est alors soumis à un fractionnement par le sulfate d'ammonium: le précipité obtenu à pH 7 et à une molarité 2 M en sulfate d'ammonium contient une part importante des protéines complexantes. Le fractionnement est poursuivi par une filtration sur Sephadex G 200. Si l'on mélange de la trypsine avec ce précipité, avant la filtration sur Sephadex, on obtient un pic d'activité estérasique en coïncidence avec les protéines exclues, un deuxième pic au niveau des protéines ayant un poids moléculaire légèrement inférieur à 100 000 et un troisième pic correspondant à la trypsine libre (fig. 1). Les fractions du premier pic sont réunies et soumises à une chromatographie sur colonne de DEAE-cellulose équilibrée contre un tampon borate 0,01 M, pH 8. L'élution est effectuée en augmentant la molarité du tampon, puis en ajoutant du NaCl. L'élution des protéines actives est obtenue avec un tampon 0,1 M, avec NaCl 0,15 M.

Ces protéines migrent en une seule bande de mobilité β-globuline en électrophorèse sur acétate de cellulose. Elles ne présentent également qu'une seule zone de faible migration en électrophorèse en gel d'acrylamide.

Leur poids moléculaire, leur caractère de précipitation par le sulfate d'ammonium, leur mobilité électrophorétique suggèrent qu'il s'agit des immunoglobulines IgA du colostrum.

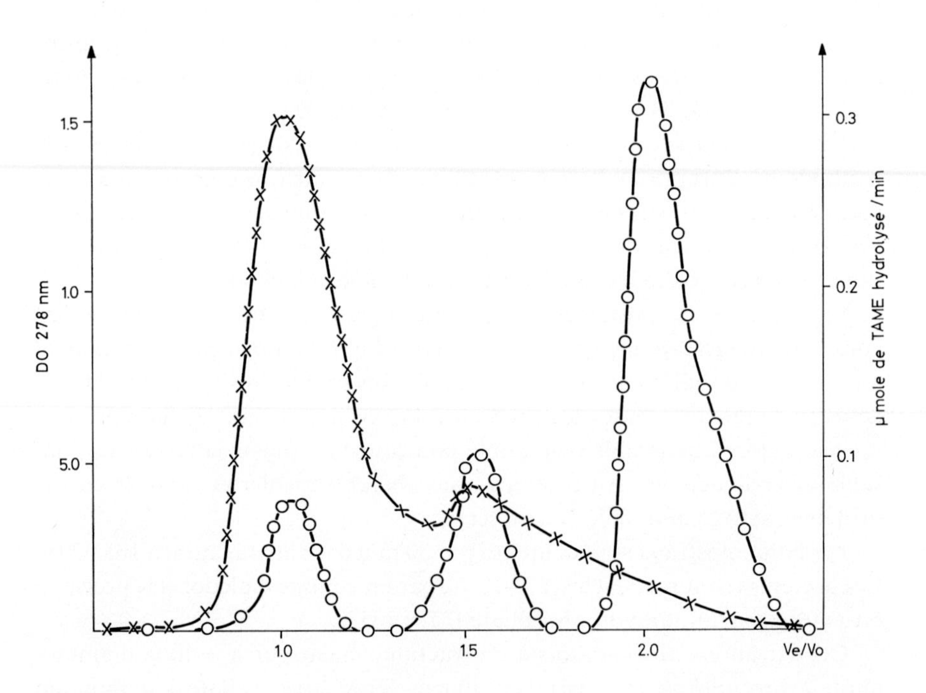

Fig. 1. Filtration sur colonne de Sephadex G 200 d'un mélange de trypsine et du précipité obtenu par le sulfate d'ammonium 2 M à pH 7. Tampon borate 0,1 M pH 8. Abscisse: rapport entre le volume d'élution (Ve) et le volume d'exclusion (Vo). Ordonnées: o–o absorbance à 278 nm (protéines); x–x μmoles de TAME hydrolysées/min.

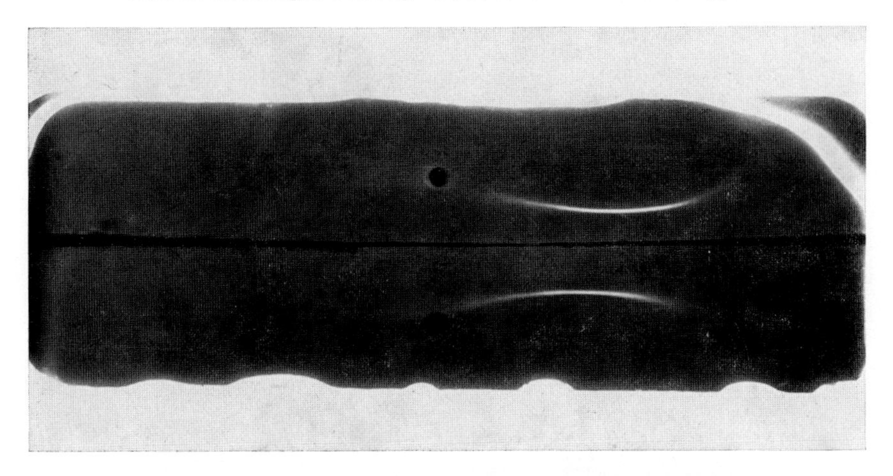

Fig. 2. Immunoélectrophorèse contre un immunsérum anti IgA. En haut, sérum humain total; en bas, IgA isolées à partir du colostrum.

Les immunoélectrophorèses effectuées en utilisant des immunsérums anti-sérum total ou anti-colostrum total ou un immunsérum spécifique anti-IgA donnent une seule ligne de précipitation caractéristique des IgA (fig. 2).

Toutefois, l'ultracentrifugation met en évidence l'hétérogénéité de cette IgA (fig. 3): le constituant principal, représentant environ 80% du total, possède une constante de sédimentation de 12,5 S; mais il apparaît également un constituant plus lourd (18,9 S) et deux constituants plus légers (6,5 S et 2,9 S). Ce résultat coïncide sensiblement avec ceux qui ont été publiés [4]. Précisons que le fait que les protéines complexant la trypsine soient exclues du Sephadex G 200 ou sédimentent au fond du tube dans l'ultracentrifugation en gradient de saccharose exclut l'intervention du constituant de faible poids moléculaire dans ce phénomène.

Divers facteurs intervenant dans la réaction d'association entre trypsine et IgA ont été étudiés.

Alors que dans le cas des α-macroglobulines, la formation du complexe semble instantanée, il faut 6 à 8 min de contact entre l'enzyme et les IgA pour que la vitesse d'hydrolyse du substrat devienne constante. L'association entre trypsine et IgA entraînant une inhibition, la vitesse d'hydrolyse du complexe est plus faible que celle de la trypsine libre et elle ne devient constante que lorsque toutes les molécules de trypsine sont associées aux IgA.

Les α-macroglobulines humaine [5], porcine [6] ou de lapin [7] peuvent lier 2 moles de trypsine par molécule, c'est-à-dire qu'elles possèdent deux

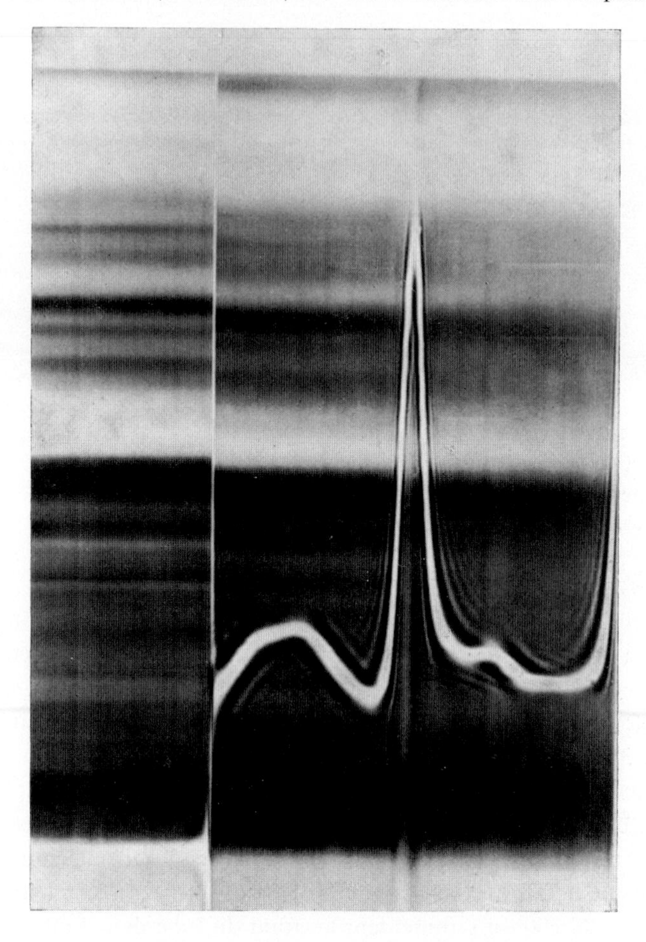

Fig. 3. Ultracentrifugation analytique des IgA du colostrum humain. 30 min de rotation à 54 000 rpm.

sites de fixation. La préparation d'IgA étudiée étant hétérogène, il n'est pas possible d'établir un rapport stoechiométrique entre trypsine et IgA. Il semble que l'inhibition maximale soit obtenue lorsque le rapport pondéral IgA/trypsine est supérieur à 30. Cependant une inhibition nette est déjà obtenue pour un rapport pondéral voisin de 1; c'est donc le constituant principal 12,5 S qui est responsable de cette propriété car les autres constituants se trouvent en trop faible proportion pour agir à un taux aussi bas.

Dans les conditions optimales de formation du complexe IgA-trypsine, c'est-à-dire après un temps de contact de 8 min à 37°C entre un poids d'IgA 30 fois supérieur à celui de trypsine, on a déterminé les variations de vitesse d'hydrolyse en fonction de la concentration en substrat: la cinétique reste michaëlienne, comme le montre la droite obtenue dans la représentation d'EADIE [8] (fig. 4). Le complexe et la trypsine libre ont pratiquement le même K_M, mais la vitesse maximale de la trypsine est nettement plus élevée que celle du complexe.

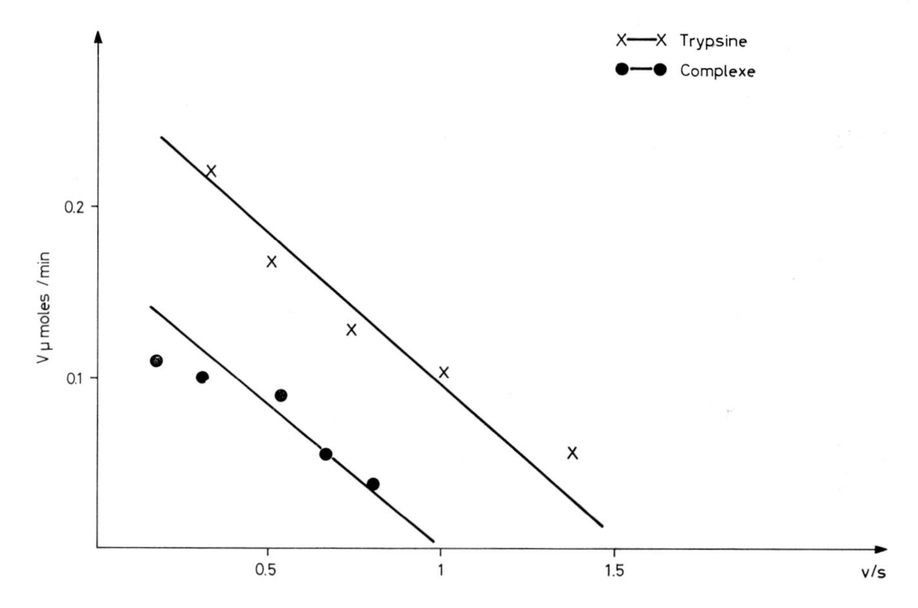

Fig. 4. Expression de la cinétique enzymatique selon la représentation d'EADIE, v = f (v/s). v = vitesse d'hydrolyse du TAME, s = concentration en TAME.

L'IgA considérée comme inhibiteur (I) se combine avec l'enzyme (E) en un site d'association différent du site catalytique. On obtient ainsi la formation d'un complexe IE qui donne avec le substrat (S) le complexe IES, avec pour résultat une diminution de la vitesse de conversion S→P (P = produits d'hydrolyse). Les courbes de la figure 4 montrent que l'on est en présence d'une inhibition non compétitive pure: l'IgA n'a aucune influence sur l'affinité de la trypsine pour le TAME, mais la vitesse catalytique du complexe est plus faible que celle de l'enzyme.

On peut concevoir que l'association entre IgA et trypsine entraîne une modification de la conformation de l'enzyme, d'où une modification du centre actif et une variation de la vitesse de catalyse.

Les immunoglobulines IgA du colostrum sont constituées par les mêmes chaînes lourdes et légères que les IgA sériques qui ont une constante de sédimentation de 7 S; mais elles se présentent essentiellement sous la forme d'un polymère obtenu à partir de trois molécules liées, d'une part entre elles par des liaisons S-S, d'autre part à une β-globuline par des liaisons non covalencielles [9]. Cette globuline, appelée dans la littérature anglaise tout d'abord 'transport piece', puis 'secretory piece', donne leur spécificité aux IgA colostrales. Elle ne se trouve pas dans les IgA sériques qui, par ailleurs, ne possèdent pas la capacité de complexer la trypsine; on peut donc supposer que cette propriété des IgA du colostrum est liée à la présence de cette β-globuline dans leur molécule polymérisée.

Summary

Ultracentrifugal experiments and gel filtration indicate that proteins of human colostrum can bind trypsin with the formation of a complex possessing enzymatic activity. One of these proteins has been isolated by a method involving ammonium sulfate precipitation, gel filtration on Sephadex G 200 and subsequent chromatography on DEAE-cellulose; by electrophoretical properties, that protein has been identified with IgA. Some characteristics of the complex have been studied.

Bibliographie

1. COUNITCHANSKY, Y.; BERTHILLIER G. et GOT R.: Mise en évidence dans le colostrum humain de protéines formant des complexes avec la trypsine et la chymotrypsine. Clin. chim. Acta *26:* 223–229 (1969).
2. BRITTEN J. and ROBERTS R. B.: High resolution density gradient sedimentation analysis. Science *131:* 32–33 (1960).
3. GOT R.: Fractionnement des protéines du lactosérum humain. Clin. chim. Acta *11:* 432–441 (1965).
4. AXELSSON H.; JOHANSSON B. G., and KRYMO L.: Isolation of Immunoglobulin A (IgA) from human colostrum. Acta chem. scand. *20:* 2,339–2,348 (1966).
5. GANROT P. O.: Interaction of plasmin and trypsin with α_2-macroglobulin. Acta chem. scand. *21:* 602–608 (1967).
6. JACQUOT-ARMAND Y.: Etude comparée de la fixation d'α_2-macroglobuline et de différents inhibiteurs sur la trypsine. C. R. Acad. Sci., Paris *264:* 2,236–2,239 (1967).

7. Berthillier G.; Got R. et Bertagnolio G.: Biochimie de l'α_1-macroglobuline de lapin. IV: Effet sur l'activité estérasique de la trypsine et de la chymotrypsine. Biochim. biophys. Acta *170:* 140–151 (1968).
8. Eadie G. S.: The inhibition of cholinesterase by physostigmine and prostigmine. J. biol. Chem. *146:* 85–93 (1942).
9. Collins-Williams C., Lamenza C., and Nizami R.: Immunoglobulin IgA: a review of literature. Ann. Aller. *27:* 225–231 (1969).

Adresse des auteurs: Dr Y. Counitchansky, Dr G. Berthillier et Dr R. Got, Laboratoire de Chimie biologique, Faculté de Médecine, *Lyon* (France)

7th int. Congr. clin. Chem., Geneva/Evian 1969; vol. 2: Clinical Enzymology, pp. 74–76
(Karger, Basel/München/Paris/New York 1970)

Isolierung eines Trypsininhibitors und zweier Trypsin-Plasmin-Inhibitoren aus den Samenblasen von Meerschweinchen

E. Fink

Institut für Klinische Chemie und Klinische Biochemie der Universität München, München

Vor vier Jahren fanden Haendle *et al.* [1] in den Samenblasen von Meerschweinchen Hemmaktivität gegenüber Trypsin. Wässrige Extrakte aus Meerschweinchen-Samenblasen enthalten je Gramm Organ eine Inhibitoraktivität gegen Trypsin von 2 000–5 000 ImU [1]. Beim Enteiweißen mit Perchlorsäure tritt ein Verlust von 15–20% auf.

Aus dem enteiweißten Extrakt wird der Inhibitor durch selektive Bindung an Trypsinharz abgetrennt. Das so gewonnene Präparat ist zu 60% rein. Abschließend wird über Sulfoäthyl-Sephadex chromatographiert, wobei eine Auftrennung in zwei Inhibitoren möglich ist. Der erste Inhibitor, M I, ist spezifisch für Trypsin, der zweite, M II, inhibiert Trypsin und Plasmin. M II tritt in zwei Komponenten (M IIa und M IIb) auf, die sich in ihrer Aminosäurezusammensetzung geringfügig unterscheiden, jedoch nicht in ihren sonstigen Eigenschaften.

Der Inhibitor M IIa ist gegenüber M IIb am N-Terminus um ein Peptid verlängert, das Serin, Prolin, Alanin und Phenylalanin enthält. Die N-terminale Gruppe ist bei M IIa Phenylalanin, bei M IIb ist sie Lysin. Der spezifische Trypsininhibitor M I hat Glutamin oder Glutaminsäure als Endgruppe. Die Molekulargewichte der Inhibitoren liegen bei 6 000.

Verwendet man beim Hemmtest BAPNA [2] als Substrat, so betragen die Hemmaktivitäten für den spezifischen Trypsininhibitor 2 700 ImU/mg bzw. 3 000 ImU/mg für den Trypsin-Plasmin-Inhibitor. Ähnliche Ergebnisse werden bei der Durchführung des Tests mit Azokasein als Substrat gefunden.

[1] 1 ImU ist die Menge Inhibitor, die 2 mU Trypsin zu 50% zu hemmen vermag.

[2] α-N-Benzoyl-DL-arginin-4-nitroanilid-hydrochlorid.

Wie sich aus den Hemmkurven (Abb. 1) berechnen läßt, beträgt die Hemmkonstante für den spezifischen Trypsininhibitor $2,3 \times 10^{-9}$ Mol/l, für den Trypsin-Plasmin-Inhibitor liegt sie unter 10^{-10} Mol/l.

Beide Inhibitoren zeigen das Phänomen der temporären Hemmung, d.h., sie werden allmählich durch Trypsin abgebaut. Wie Abb. 2 zeigt, wird der spezifische Trypsininhibitor M I wesentlich rascher inaktiviert als der Trypsin-Plasmin-Inhibitor M II, was auf Grund der um etwa zwei Zehnerpotenzen höheren Hemmkonstanten verständlich ist.

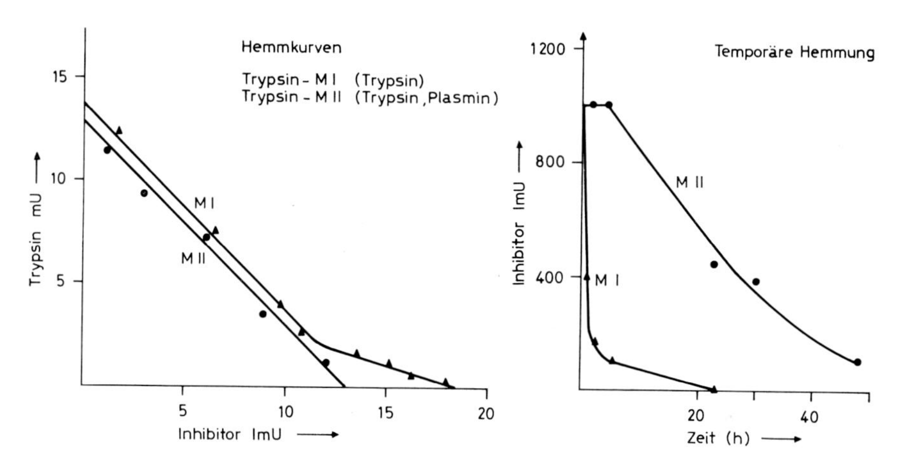

Abb. 1. Hemmkurven des spezifischen Trypsininhibitors M I und des Trypsin-Plasmin-Inhibitors M II in Gegenwart von BAPNA. Es wurde eine definierte Menge Trypsin vorgelegt und nach Zugabe variierender Inhibitormengen (Abszisse) die tryptische Restaktivität (Ordinate) gemessen.

Abb. 2. Temporäre Hemmung des spezifischen Trypsininhibitors M I und des Trypsin-Plasmin-Inhibitors M II. Die Inhibitoren wurden mit einem geringen Überschuß an Trypsin bei 37° C inkubiert. In Abständen wurden Proben entnommen und die noch vorhandene Inhibitoraktivität nach Ausfällen des Trypsins mit Perchlorsäure gemessen (Ordinate).

Im Gegensatz zu anderen Trypsininhibitoren wird der spezifische Trypsininhibitor aus Meerschweinchensamenblasen durch Plasmin rasch zerstört. Auch der Trypsin-Plasmin-Inhibitor wird durch Plasmin abgebaut, jedoch wesentlich langsamer.

Wird der Trypsin-Plasmin-Inhibitor mit einem Überschuß an Plasmin inkubiert, so verliert er seine Hemmwirkung gegen Trypsin. Möglicherweise ist also für die Aktivität gegen beide Enzyme dasselbe Hemmzentrum verantwortlich. Nach Acylierung mit Maleinsäureanhydrid [2] sind beide Inhibitoren inaktiv, d.h., beide haben einen Lysinrest im reaktiven Zentrum[3]; der acylierte Trypsin-Plasmin-Inhibitor ist auch gegen Plasmin inaktiv, was wiederum für ein einziges Hemmzentrum für beide Enzyme spricht. In saurer Lösung wird die Aktivität beider Inhibitoren quantitativ regeneriert.

Bei allen bisher untersuchten Säugetieren und beim Menschen findet man im Sperma bzw. in den Samenblasen Inhibitoraktivität gegenüber Trypsin, das Meerschweinchen ist also kein Sonderfall.

Summary

Two proteinase inhibitors (molecular weight near 6,000) were isolated from seminal vesicles of guinea pigs by use of water insoluble trypsin resin and chromatography on Sulphoethyl Sephadex. Inhibitor I inhibits only trypsin ($K_i = 2{,}3 \times 10^{-9}$); inhibitor II, trypsin ($K_i < 10^{-10}$ Mol/l), and plasmin. Both inhibitors show temporary inhibition. From inhibitor II a modified form was isolated as well.

Literatur

1. HAENDLE, H.; FRITZ, H.; TRAUTSCHOLD, I. und WERLE, E.: Über einen hormonabhängigen Inhibitor für proteolytische Enzyme in männlichen accessorischen Geschlechtsdrüsen und im Sperma. Z. physiol. Chem. *343:* 185 (1965).
2. FRITZ, H.; GEBHARDT, M.; HOCHSTRASSER, K. und WERLE, E.: Identifizierung von Lysin- und Argininresten als Hemmzentren von Proteaseinhibitoren mit Hilfe von Maleinsäureanhydrid und Butandion-(2–3), Z. physiol. Chem. *350:* 933 (1969).

[3] Anmerkung bei den Korrekturen: Dieser Befund ist für den Inhibitor M I aufgrund neuer Ergebnisse überholt, siehe E. FINK, Dissertation, Universität München, 1970.

Adresse des Autors: Dr. E. FINK, Institut für Klinische Chemie und Klinische Biochemie der Universität München, *D-8 München* (Deutschland).

7th int. Congr. clin. Chem., Geneva/Evian 1969; vol. 2: Clinical Enzymology, pp. 77–81
(Karger, Basel/München/Paris/New York 1970)

Corrélation entre la capacité d'inhibition protéasique et le taux d'α_1-antitrypsine et d'α_2-macroglobuline des liquides d'ascite et pleuraux

J. BIETH, F. MIESCH et P. MÉTAIS
Laboratoire de la Clinique Médicale A, Hospices Civils, Strasbourg

Le sérum humain contient plusieurs globulines capables d'inhiber plus ou moins spécifiquement la trypsine ou la chymotrypsine. Notre intérêt s'est porté récemment sur l'étude du pouvoir inhibiteur de protéases, de liquides extravasculaires comme le liquide céphalorachidien [1] ou les liquides pleuraux ou d'ascite.

Pour déterminer ces protéines, deux types de techniques ont été utilisées. L'emploi d'antisérums spécifiques de l'α_1-antitrypsine, de l'α-$_2$-macroglobuline et de l'inter-α-inhibiteur (protéine π) qui nous a été remis gracieusement par Madame STEINBUCH, nous a permis de démontrer la présence de ces inhibiteurs dans les divers liquides. A l'aide de la technique d'électro-immuno-diffusion de LAURELL [3] employant ces divers antisérums spécifiques, il nous a été possible de déterminer le taux de chacun de ces inhibiteurs et particulièrement de l'α_1-antitrypsine et de l'α_2-macroglobuline dans les liquides pleuraux et d'ascite prélevés au cours de maladies diverses (cirrhoses, cancers, troubles cardiaques et pancréatiques). Il est alors facile de comparer ces résultats aux mesures enzymatiques de capacité d'inhibition trypsique ou chymotrypsique déterminées à l'aide de techniques que nous avons décrites précédemment pour le sérum et le liquide céphalo-rachidien [2] (tableau I).

Méthodes

L'étude a porté sur 43 épanchements: quatre déterminations sont effectuées sur chacun d'entre eux. La mesure des capacités d'inhibition protéasique vis-à-vis de la trypsine ou de la chymotrypsine s'effectue ainsi: après réaction en présence d'un excès d'enzyme, cet excès non inhibé est dosé grâce à des substrats chromogènes: le benzoylarginine-p. nitroanilide (BAPNA) pour la trypsine et le succinyl-phénylalanine-p.nitroanilide

Tableau I. Les inhibiteurs de trypsine et de chymotrypsine de quelques liquides biologiques

	Enzymes inhibées		Se trouvant dans			
	Trypsine	Chymo-trypsine	Sérum	Liq. céphalo-rachidien	Liq. d'ascite	Liq. pleural
α_1-antitrypsine (α_1 3,5 S glycoprotéine)	+	+	+	+	+	+
α_1-X-glycoprotéine (α_1-antichymotrypsine)	—	+	+	?	?	?
Inter-α-inhibiteur (protéine π)	+	+	+	+	+	+
α_2-macroglobuline	+*	+*	+	+	+	+

* Seulement quand l'activité enzymatique est vérifiée au moyen d'un substrat protéique.

(SUPHEPA) pour la chymotrypsine. Les dosages immunochimiques d'α_1-antitrypsine et d'α_2-macroglobuline sont réalisés par la méthode de LAURELL [3] qui consiste à faire migrer électrophorétiquement un antigène dans un gel d'agarose contenant des anticorps. Le précipité antigène-anticorps, révélé par le bleu de Coomassie prend un aspect de pic acuminé dont la hauteur est proportionnelle à la concentration de l'antigène.

Résultats

La distribution des valeurs obtenues présente de grandes similitudes entre les liquides pleuraux et d'ascite. La distribution des chiffres obtenus pour la capacité d'inhibition trypsique se situe entre 0,3 unités et les valeurs du sérum normal, soit 4,5 à 6,5 unités: la moyenne est à 3,0 unités pour l'ascite et 3,2 pour le liquide pleural, soit autour de 60% de la capacité sérique.

La capacité d'inhibition chymotrypsique est dans ces liquides comprise entre 0,02 et 0,50 unités avec une moyenne de 0,18 pour l'ascite et de 0,21 pour le liquide pleural. Là encore, la capacité d'inhibition chymotrypsique de ces liquides est environ 55% de celle du sérum.

Le dosage immunochimique de l'α_1-antitrypsine fournit les résultats directement en pourcentage par rapport à un sérum normal. Ils sont compris entre 10 et 170% avec une moyenne de 66% pour l'ascite et 71% pour le liquide pleural.

Enfin le taux d'α_2-macroglobuline est toujours le plus faible. Il est compris entre 5 et 66% par rapport à un sérum normal et la moyenne se situe à 21% pour l'ascite et 25% pour le liquide pleural.

La répartition des valeurs obtenues en fonction de l'origine étiologique du liquide montre que les valeurs de capacités d'inhibition et d'α_1-antitrypsine sont les plus élevées au cours des processus cancéreux, puisque 18 résultats sur 32 dosages effectués chez 8 malades fournissent des résultats supérieurs au sérum normal. Au contraire 53 valeurs sur 64 dosages effectués chez 16 cirrhotiques et 40 valeurs sur 60 dosages effectués chez 15 insuffisants cardiaques sont inférieures à la norme sérique. Enfin chez les 4 pancréatiques étudiés, l'ensemble des résultats se répartit autour de la norme du sérum, mais on constate toujours dans le liquide de ces sujets une élévation de l'α_1-antitrypsine supérieure à celle des capacités d'inhibition.

La comparaison entre le sérum, prélevé au même moment, et le liquide extravasculaire du même sujet (fig. 1) met en évidence, dans un petit nombre de cas, la richesse toujours plus grande du sérum en inhibiteurs par rapport aux liquides. La différence est particulièrement nette pour l'α_2-macroglobuline.

Enfin la corrélation entre les taux d'α_1-antitrypsine et d'α_2-macroglobuline n'est pas significative (fig. 2). Si l'on admet que ces protéines franchissent passivement la barrière capillaire à la suite de diverses anomalies pathologiques, on peut mettre ce manque de corrélation sur le compte des diffé-

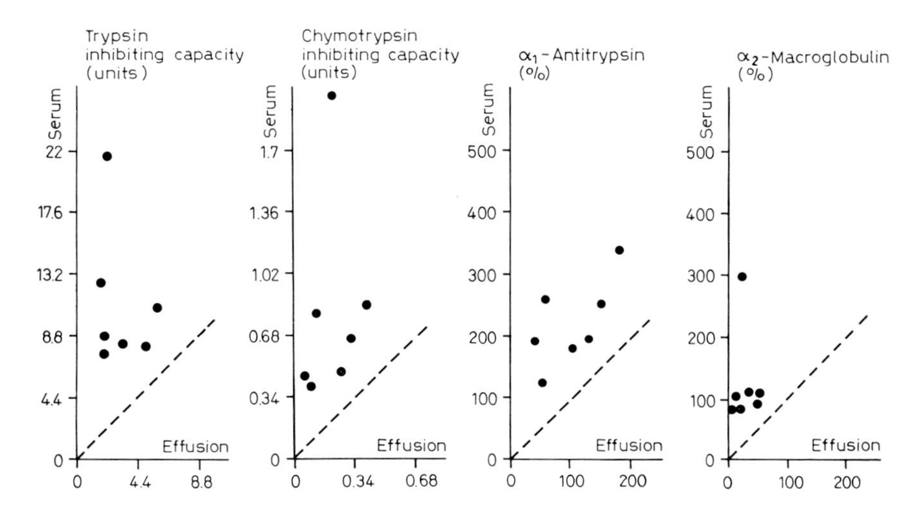

Fig. 1. Comparaison entre sérum et liquides pleuraux ou d'ascite (7 malades).

rences de masse moléculaire existant entre ces deux protéines. Au contraire, la bonne corrélation entre la capacité d'inhibition protéasique laisse supposer que l'inhibition de la trypsine et de la chymotrypsine est due aux mêmes substances. Que cette substance soit en général l'α_1-antitrypsine semble être démontré par la bonne corrélation constatée entre le dosage de cette protéine et les pouvoirs inhibiteurs. On remarque cependant que dans le cas de la corrélation entre l'inhibition chymotrypsique et l'α_1-antitrypsine, les résultats sont moins groupés: ce fait serait en accord avec la présence d'autres inhibiteurs de la chymotrypsine dans ces liquides.

Enfin dans un certain nombre de dosages, le taux d'α_1-antitrypsine, dosé par une méthode immunochimique, est nettement plus élevé que la capacité d'inhibition. On peut en conclure à la présence d'α_1-antitrypsine immunologiquement réactive, mais incapable de se combiner aux protéases. Il est à noter que la plupart de ces résultats sont obtenus chez des sujets souffrant de pancréatites.

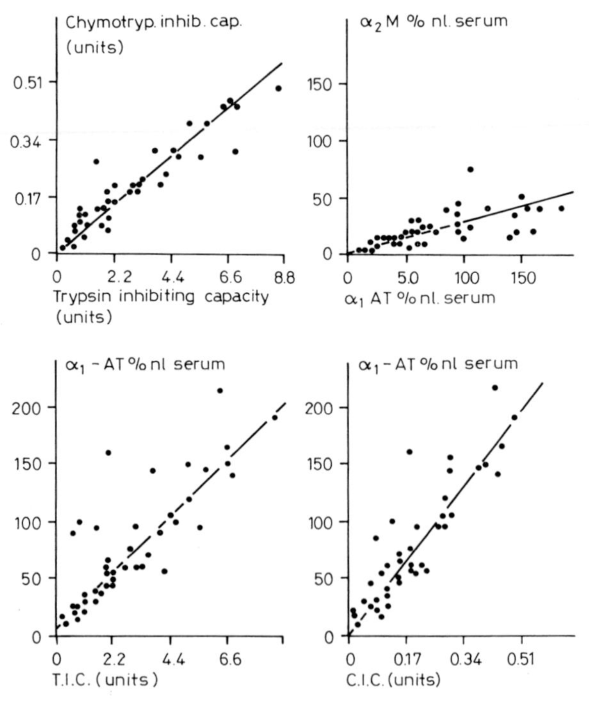

Fig. 2. Corrélation entre la capacité d'inhibition de trypsine et de chymotrypsine et le taux de l'α_1-antitrypsine et de l'α_2-macroglobuline dans les liquides pleuraux et d'ascite.

Une étude plus détaillée de ces liquides est en cours: la gel-filtration sur Sephadex G 200 a permis de confirmer la présence d'α_2-macroglobuline liée à de la trypsine et celle d'α_1-antitrypsine biologiquement inactive. Des résultats préliminaires nous portent à penser que cette α_1-antitrypsine représente l'inhibiteur lié aux protéases endogènes.

Conclusion

L'α_1-antitrypsine est la globuline responsable de l'inhibition protéasique des liquides d'effusion. Ce fait est en accord avec ce qui a été montré pour le sérum et le liquide céphalo-rachidien, mais le rôle de l'α_1-antitrypsine et de l'α_2-macroglobuline semble être plus précis dans ces liquides que dans le sérum. Ces globulines doivent être les inhibiteurs essentiels des protéases lors des inflammations et en particulier au cours des anomalies pancréatiques où il semble possible de mettre en évidence une α_1-antitrypsine biologiquement inactive.

Summary

The trypsin- and chymotrypsin-inhibiting capacity (TIC and CIC) and the immunochemically determined α_1-antitrypsin (α_1AT), and α_2-macroglobulin (α_2M) content have been compared in the pleural and ascitic fluids from 43 patients suffering from various diseases. The following results were obtained:

1. The mean values of TIC, CIC and α_1AT are around 60% of normal human serum, whereas those of α_2M are much lower (a. 20%).
2. The highest values of TIC, CIC and α_1AT are found in carcinoma, whereas the lowest values are found in liver cirrhosis. Cardiac insufficiency or pancreatic diseases give medium values.
3. The α_1AT content usually parallels the protease inhibiting capacity except in pancreatic diseases where endogenous proteases may be bound to α_1AT.

Bibliographie

1. Bieth, J.; Miesch, F. et Métais, P.: Clin. chim. Acta *24:* 203 (1969).
2. Bieth, J.; Métais, P. et Warter, J.: Enzymol. biol. clin. *10:* 243–257 (1969).
3. Laurell, C. B. Anal. Biochem. *15:* 45 (1966).

Adresse des auteurs: Dr. J. Bieth, Dr. F. Miesch et Dr. P. Métais, Laboratoire de la Clinique Médicale A, Hospices Civils, *Strasbourg-67* (France)

Communications
a) Methods of Enzyme Determination

7th int. Congr. clin. Chem., Geneva/Evian 1969; vol. 2: Clinical Enzymology, pp. 82–88 (Karger, Basel/München/Paris/New York 1970)

Etude de la conservation du sang capillaire prélevé sur papier pour le dosage de la galactotransférase et de la glucose-6-phosphate deshydrogénase

C. Dorche, C. Kissin[1], C. Collombel, M. Mathieu[2], M. O. Rolland et J. Cotte

A.R.D.E.M.M.E., Laboratoire de Biochimie, Hôpital d'Enfants Debrousse, Lyon

Dans le cadre d'un programme de dépistage systématique des maladies métaboliques du nouveau-né, nous avons d'abord recherché la phénylcétonurie, puis nous y avons ajouté le dépistage de la galactosémie et du déficit en glucose-6-phosphate deshydrogénase (G-6-PD).

Avant d'entreprendre ces dépistages, nous avons étudié la stabilité du prélèvement sur papier en examinant la conservation des activités de la galactotransférase et de la G-6-PD sur du sang prélevé sur papier et conservé pendant un mois. Nous avons prélevé le sang capillaire de 5 sujets normaux et nous l'avons posé sur des papiers à prélèvements numérotés de 1 à 5. Le sang n° 1 a été conservé à +4°, les autres à la température du laboratoire. Tous les jours, nous avons évalué qualitativement et quantitativement l'activité de ces 2 enzymes dans les 5 sangs.

Méthode

Nous avons utilisé les spot tests de Beutler [1, 2].

Principe (fig. 1)

La galactotransférase catalyse en présence de l'uridyl-diphosphoglucose (UDPG) la transformation du galactose-1-phosphate (Gal-1-P) en glucose-1-phosphate (G-1-P).

[1] Attachée de recherche I.N.S.E.R.M.

[2] Chargée de recherche I.N.S.E.R.M.

Celui-ci, sous l'influence de la phosphoglucomutase puis de la phospho-hexose isomérase, donne du β-glucose-6-phosphate (β-G-6-P). En présence de NADP et de la G-6-PD le β-G-6-P se transforme en 6-phosphogluconate et le NADP est réduit en NADPH + H^+ qui est fluorescent sous lumière ultra-violette. La réaction se poursuit sous l'influence de la 6-phosphogluconate deshydrogénase et entraîne la réduction d'une autre molécule de NADP.

Si l'enzyme étudié est présent, la réaction se déroule et le NADP est réduit.

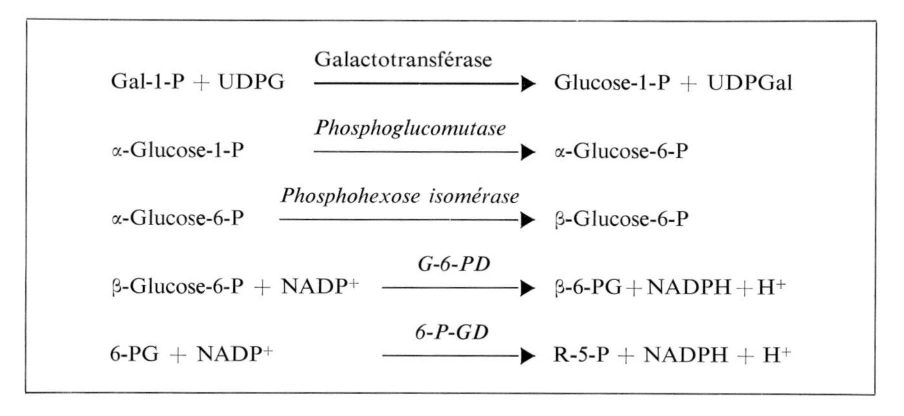

Fig. 1. Schéma de la réaction. PGM = phosphoglucomutase, PHI = phosphohexose isomérase, G-6-PD = glucose-6-phosphate deshydrogénase, 6-PGD = 6-phosphogluconate deshydrogénase.

Réalisation qualitative pour la galactotransférase [2]

On met en présence en milieu tamponné le substrat qui apporte UDPG + Gal-1-P + NADP et la tache de sang découpée dans la carte à prélèvement qui contient les enzymes. Après 3 h d'incubation à 37° C, on reprend une partie du mélange que l'on dépose sur un papier. Celui-ci est ensuite examiné en fluorescence sous une lumière ultra-violette (fig. 2). Les sangs normaux donnent des taches fluorescentes, les déficients des taches sombres.

Réalisation qualitative pour la G6PD [1]

La technique est la même mais le substrat apporte G-6-P + NADP et l'incubation n'est que de 15 min. à 22° C.

Modification quantitative pour la galactotransférase [3]

La réaction est la même que pour le dépistage qualitatif mais après 30 min d'incubation à 37° C, on reprend 20 μl du mélange qui sont mis en milieu tamponné et passés dans un fluorimètre 'Turner' contre une gamme de NADPH (fig. 3).

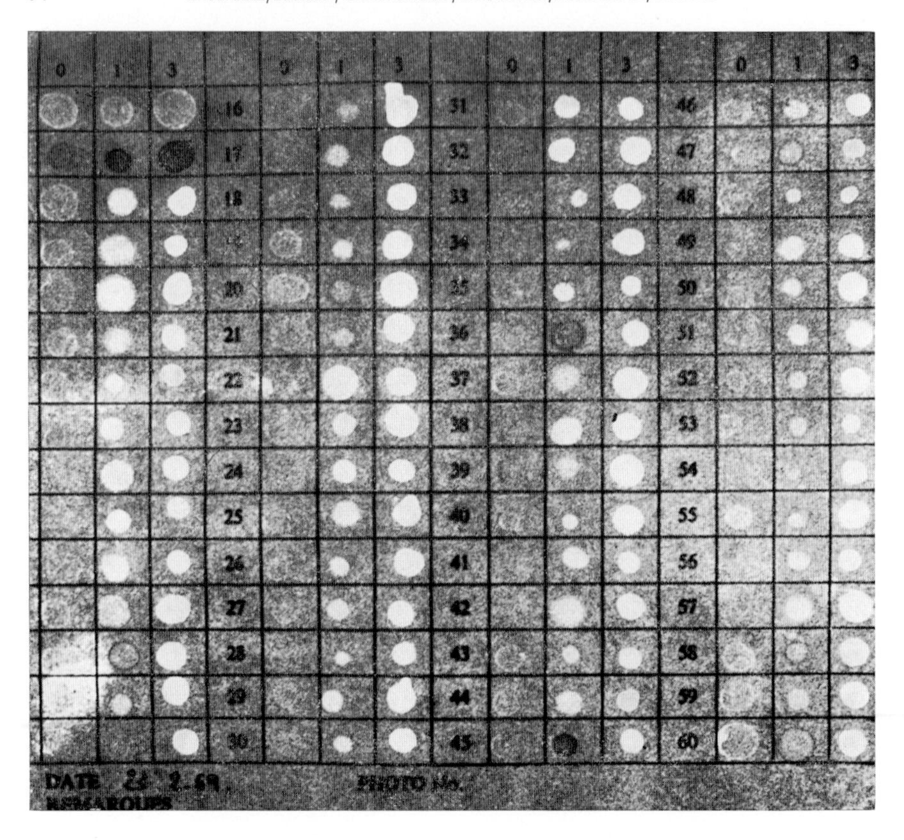

Fig. 2. Représentation d'une série de dépistage de la galactosémie. Les chiffres horizontaux donnent les heures d'incubation. Les chiffres verticaux donnent les numéros des échantillons: n° 1: témoin réactif; n° 2: témoin galactosémique.

Les résultats sont ensuite rapportés au taux d'hémoglobine contenu dans la tache de sang étudiée et à l'unité de temps.

Nous avons évalué de manière similaire la G-6-PD.

Résultats

Qualitatifs

Pour la galactotransférase (fig. 4), on voit que le sang n° 1 qui a été conservé à +4°C reste plus longtemps fluorescent. Pour les autres, la fluorescence

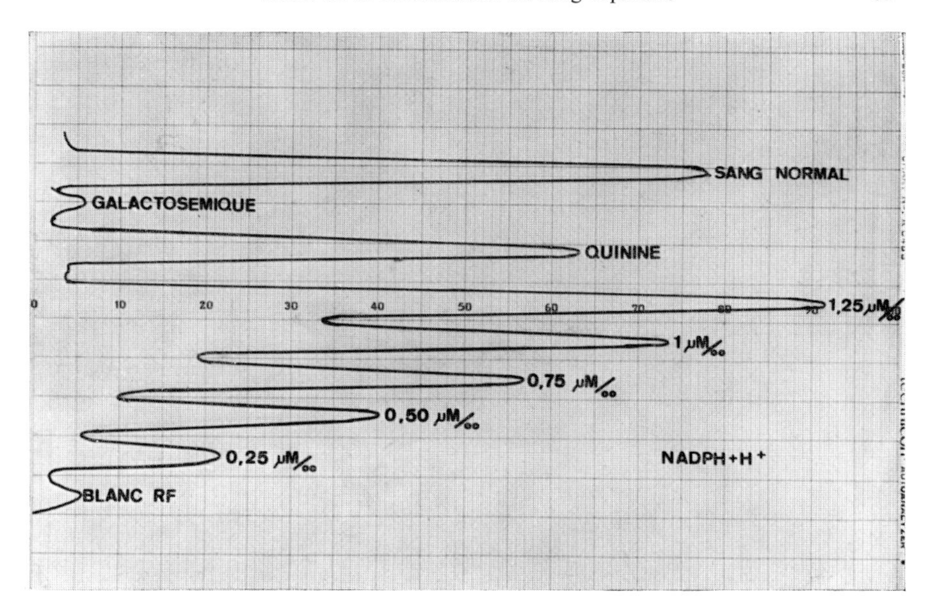

Fig. 3. Etalonnage d'un fluorimètre «Turner» avec une gamme de NADPH. Exemple de la fluorescence produite par un sang normal et par un sang de déficient (égale à celle du blanc réactif: aucune réduction du NADP n'a eu lieu).

reste visible jusqu'au 21e jour, puis diminue beaucoup et devient à peine visible un mois après le prélèvement.

Pour la G-6-PD (fig. 5), la différence entre le sang conservé à +4°C et à +22° C est moins grande. La fluorescence au 21e jour est plus nettement visible que pour la galactotransférase.

Quantitatifs

Pour la galactotransférase (fig. 6) aussi, on voit la différence entre le sang conservé à +4°C et à +22°C. A +4°C, en une semaine, il y a eu une perte de 38% d'activité, en un mois de 50%. A +22°C, en une semaine, il y a eu une perte de 55% d'activité et en un mois de 75%.

On voit aussi qu'il y a une chute très rapide de l'activité dans les premiers jours de conservation puis moins rapide par la suite.

Pour la G-6-PD (fig. 7), à +4°C en une semaine on perd 28% d'activité, en un mois 55%. A +22°C, en une semaine, on perd 52% d'activité et en un mois 70%.

4

5

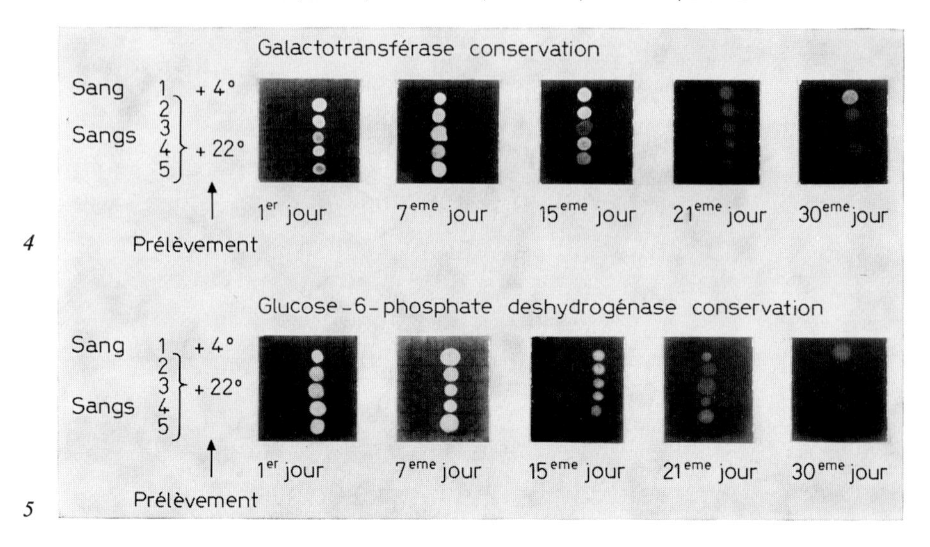

Fig. 4. Représentation de la fluorescence donnée par les 5 sangs par dosage de la galactotransférase 1, 7, 15, 21, 30 jours après le prélèvement.

Fig. 5. Représentation de la fluorescence donnée par les 5 angs par dosage de la G-6-PD 1, 7, 15, 21, 30 jours après le prélèvement.

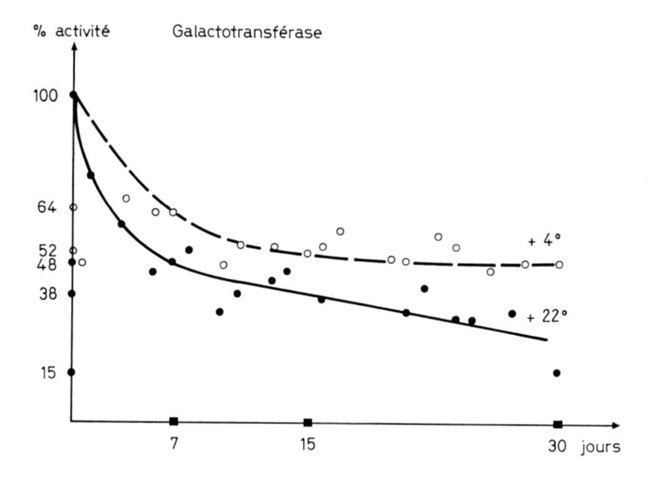

Fig. 6. Pourcentage de conservation de l'activité de la galactotransférase en fonction du temps.

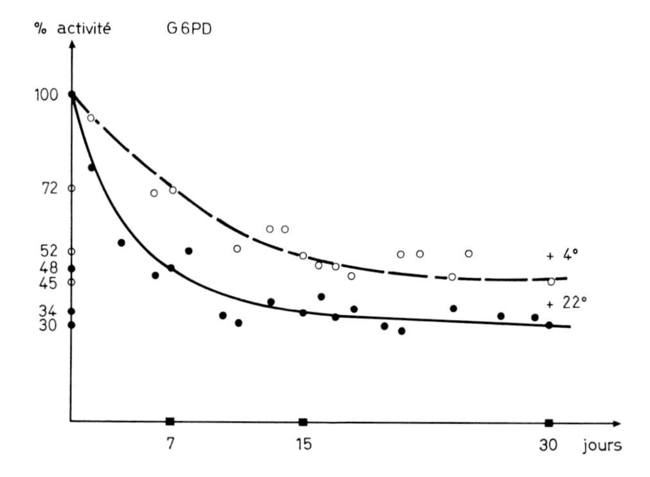

Fig. 7. Pourcentage de conservation de l'activité de la G-6-PD en fonction du temps.

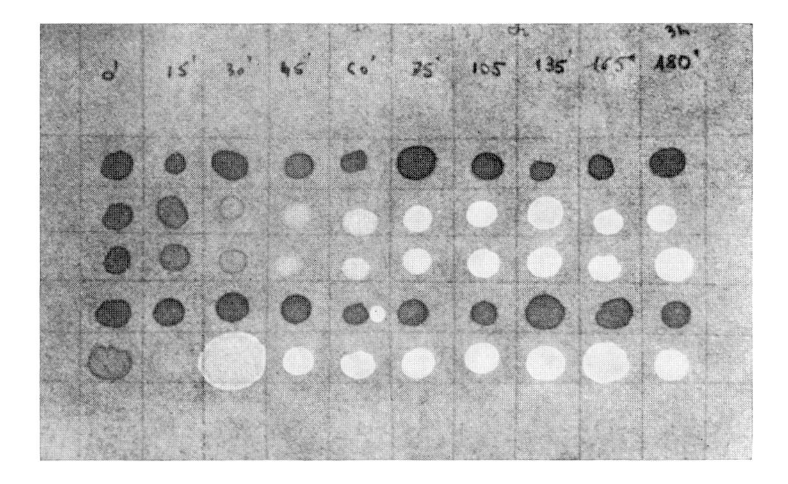

Fig. 8. Recherche d'hétérozygotisme chez les parents d'un enfant galactosémique. Les chiffres horizontaux donnent les temps d'incubation en minutes, la première ligne provient d'un sang témoin galactosémique, la 2e et la 3e ligne du sang des parents, la 4e ligne de l'enfant galactosémique et la 5e ligne d'un sang normal.

De la comparaison de ces graphiques, on peut tirer la conclusion que la G-6-PD se conserve au moins aussi bien que la galactotransférase, contrairement à ce que l'on croyait jusqu'à ces derniers temps.

En comparant les photos de fluorescence qualitative et le taux de l'enzyme au même jour, on voit que la fluorescence reste visible qualitativement même lorsque l'activité est tombée en-dessous de 50%. Ceci exclut la possibilité de dépister systématiquement les hétérozygotes par cette méthode. Mais en pratiquant, le jour du prélèvement, comparativement à un sang normal des spot tests à différents temps d'incubation, on arrive très bien à les dépister [1] (fig. 8). On constate un retard dans l'apparition de la fluorescence pour le sang des parents alors qu'après 3 h d'incubation la fluorescence est comparable à celle du témoin.

Par ces essais, nous avons montré qu'un dépistage de la galactosémie et du déficit en G-6-PD était possible sur du sang prélevé sur papier, puisque la fluorescence persiste de manière visible pendant 3 semaines. Or, pour être efficace, un dépistage, et surtout celui de la galactosémie, doit se situer bien en-deçà de cette limite.

Summary

The authors have studied the stability of capillary blood, dried on filter paper in the estimation of galactose-1-phosphate, uridyl transférase, and glucose-6-phosphate dehydrogenase activity. The measurements have been performed according to the BEUTLER's technique which is based on reduction of NADP. The results show that, qualitatively, fluorescence remains clear for 3 weeks, while, quantitatively, the activity decreases quickly for 1 week and more slowly for 1 month.

Bibliographie

1. BEUTLER, E.: A series of new screening procedures for pyruvate kinase deficiency, glucose-6-phosphate dehydrogenase deficiency, and glutathione reductase deficiency. Blood *28:* 553–561 (1966).
2. BEUTLER, E. and BALUDA, M.: A simple spot screening test for galactosemia, J. Lab. clin. Med. *68:* 137–141 (1966).
3. BEUTLER, E. and MITCHELL, M.: New rapid method for the estimation of red cell galactose-I-phosphate uridyl transferase activity. J. Lab. clin. Med. *72:* 527–532 (1968).

Adresse des auteurs: Dr. C. DORCHE, Dr. C. KISSIN, Dr. C. COLLOMBEL, Dr. M. MATHIEU, Dr. M. O. ROLLAND et Dr. J. COTTE, A.R.D.E.M.M.E., Laboratoire de Biochimie, Hôpital d'Enfants Debrousse, 29, rue Sœur-Bouvier, *69 - Lyon 5e* (France)

7th int. Congr. clin. Chem., Geneva/Evian 1969; vol. 2: Clinical Enzymology, pp. 89–94
(Karger, Basel/München/Paris/New York 1970)

Dependency of the Glutathione Reductase Activity on the Riboflavin Status

D. GLATZLE
Department of Vitamin and Nutritional Research, F. Hoffmann-La Roche & Co. Ltd., Basle

The $NADPH_2$-dependent glutathione reductase in erythrocytes is a flavoenzyme where the prosthetic group behaves like flavin adenine dinucleotide (FAD).

If the amount of the coenzyme FAD is insufficient, the activity of this enzyme might be impaired. In this case, addition of FAD could stimulate the activity, if the apoenzyme has been synthesized despite a shortage of FAD.

A suboptimal supply of the precursor riboflavin may be the reason for this deficiency of FAD which could affect the activity of the $NADPH_2$-dependent glutathione reductase.

To check this, we use an enzymic assay based on a modified spectrophotometric method of RACKER [1], as shown in figure 1. The enzyme activity of hemolyzed erythrocytes is measured in two cuvettes, one of which contains additional FAD. The consumption of $NADPH_2$ in both cuvettes is compared following the decrease of absorbance at about 340 nm. The reaction rate without FAD-addition is used as reference and an activation coefficient α can be calculated showing the relative stimulation by FAD.

The assay mixture is buffered using 0.1 M potassium phosphate of pH 7.4. Washed erythrocytes are hemolyzed by a 20-fold dilution with water. Freshly prepared solutions of the tetrasodium salt of $NADPH_2$ and the monosodium salt of the FAD from Boehringer, Mannheim, are used.

After a preincubation period at 35° C, when the optical density of the assay does not change at 334 nm, the substrate, oxidized glutathione, is added and the oxidation of reduced nicotinamide adenine dinucleotide phosphate in the cuvettes is measured at the same temperature for the first 10 min. The activation coefficient α is calculated using the following equation:

$$GSSG + NADPH_2 \underset{}{\overset{EGR\ (FAD)}{\rightleftharpoons}} 2\ GSH + NADP$$

assay mixture (in 1.8 ml)	contents per ml assay cuvette 1	cuvette 2
buffer 0.1 M potassium phosphate pH 7.4	7.5×10^{-5} mole	7.8×10^{-5} mole
hemolysate (washed erythrocytes diluted 1:20 with water)	0.1 ml	0.1 ml
$NADPH_2$	10^{-7} mole	10^{-7} mole
$EDTA\text{-}K_2$	2.3×10^{-6} mole	2.3×10^{-6} mole
FAD	8×10^{-9} mole	0
after preincubation at 35° C for 5 minutes addition of		
GSSG	2×10^{-7} mole	2×10^{-7} mole

enzyme activity determined by measuring the decrease in absorbance ΔA at 334 mμ

$$\text{activation coefficient } \alpha = \frac{\text{diminution of } NADPH_2 \text{ in cuvette 1}}{\text{diminution of } NADPH_2 \text{ in cuvette 2}} = \frac{\Delta A_1}{\Delta A_2}$$

Fig. 1. Glutathione reductase test procedure.

$$\text{activation coefficient } \alpha = \frac{\text{diminution of } NADPH_2 \text{ in cuvette 1}}{\text{diminution of } NADPH_2 \text{ in cuvette 2}}.$$

In figure 2, these activation coefficients are compiled for 36 blood donors. We generally find a slight activation or inhibition of the enzyme by FAD. The inhibition might be due to the assay system we use. It is subject to the purity of FAD and to its concentration. Two samples show higher activation by FAD.

Elevated activation indices are found also in blood samples we received from Japan (fig. 3), which are highly suspect of riboflavin deficiency. They were supplied to us by the generosity of Prof. ARAKAWA (Tohoku University).

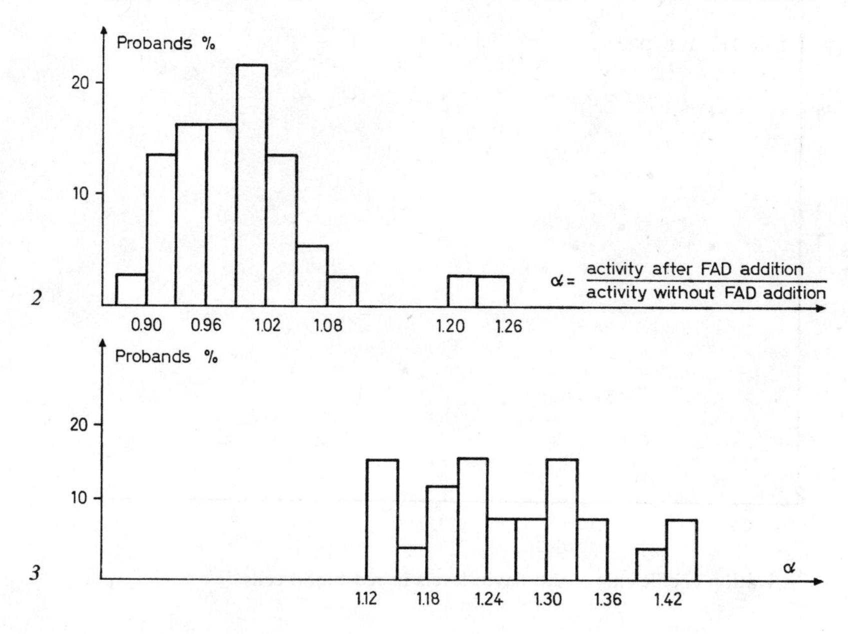

Fig. 2. Activation coefficients α from blood donors.
Fig. 3. Activation coefficients α from Japanese school children.

As we mentioned at the beginning of this communication, a stimulation by FAD expressed by high activation coefficients is expected for riboflavin deficiency. The low values in figure 2 for most of the blood donors suggest that there was no shortage in flavin adenine dinucleotide supply for the glutathione reductase system. Further investigations with about 190 blood donors lead to the assumption that in our test system activation coefficients of $\alpha \geq 1.20$ could be interpreted as signs for a biochemical riboflavin deficiency.

In the groups checked so far it was statistically significant that there is a negative correlation between flavin levels in red blood cells estimated by a fluorimetric method and the glutathione reductase activation coefficient α, which is not always as evident as in figure 4. There is a trend for low activation coefficients when high flavin levels are found in the erythrocytes and *vice versa.*

The riboflavin contents of red blood cells are estimated according to a modified method of BURCH *et al.* [2] in extracts from red blood cell hemo-

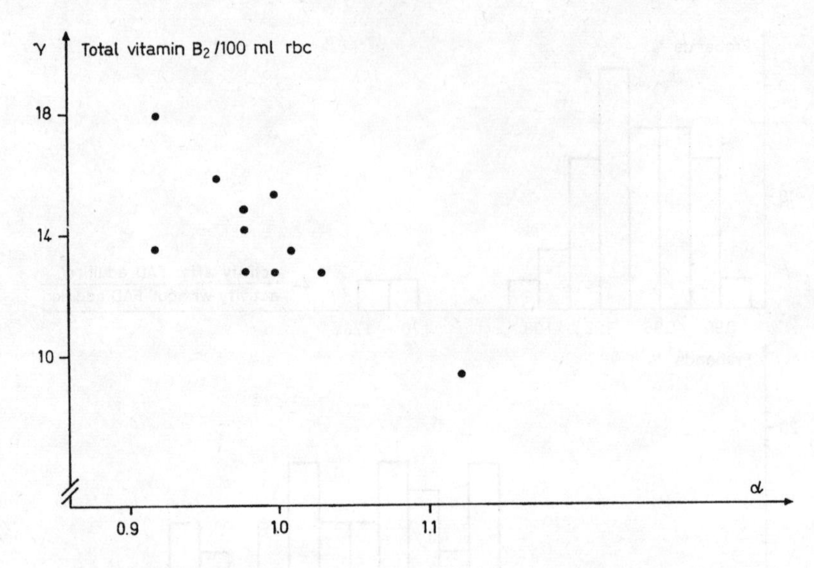

Fig. 4. Activation coefficients α and flavin levels in red blood cells.

lysate which do not contain, for instance, the flavin component of succinic dehydrogenase. Therefore, the values showing the riboflavin equivalents in 10^{-6} g per 100 ml rbc may be too low. They proved to be useful, however, as a parameter to check relationships.

In figure 5, data from geriatric patients are added as dark points to those from adults of a holiday resort we have looked at just now and which are presented here as open circles. Especially when samples showing higher activation coefficients are checked, the correlation although still significant is not so evident any more, which might indicate that a 'normal' cellular flavin level does not necessarily mean that a special biochemical function influenced by the coenzyme is already optimal.

Figure 6 shows that the activation coefficient α is shifted to lower values by increasing the oral riboflavin supply in the same persons. The same behavior was found in a larger group of geriatric patients. 10 mg additional riboflavin was administered for 7 days. The values of 'normal' people stay in a region which is found in other healthy people, too. Abnormally high values shift to normal, as can be seen in this case (1.3 → 0.9).

We therefore suggest to try this test as a parameter for the detection of riboflavin deficiency.

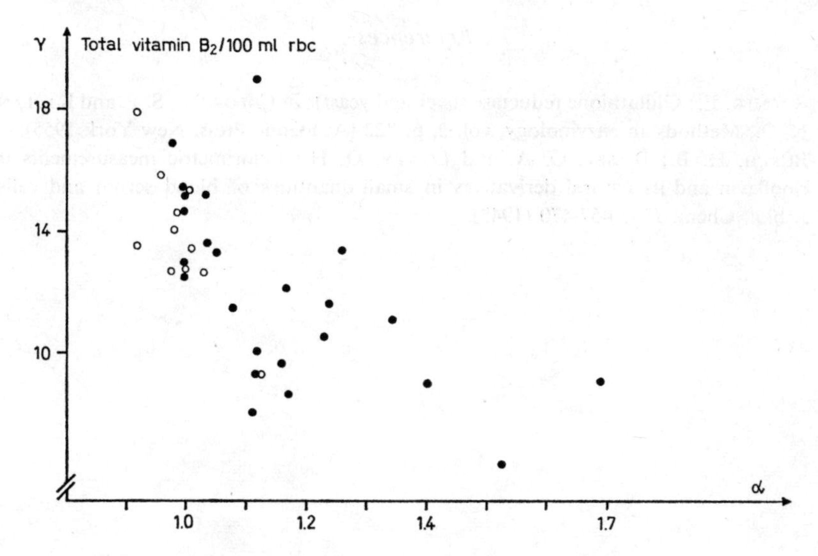

Fig. 5. Activation coefficients α and flavin levels in red blood cells.

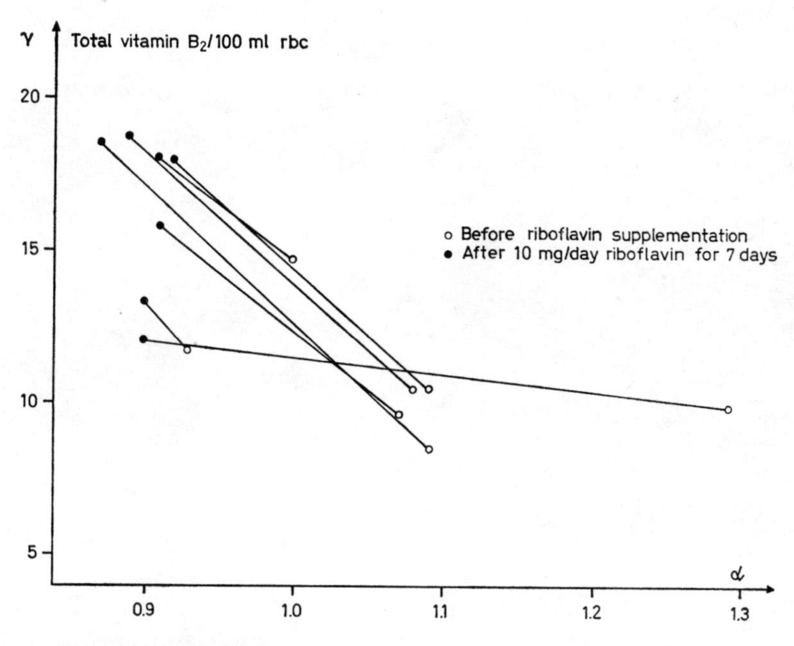

Fig. 6. Influence of riboflavin supplementation on the activation coefficient and the flavin levels in red blood cells.

References

1. RACKER, E.: Glutathione reductase (liver and yeast); in COLOWICK, S. P. and KAPLAN, N. O., Methods in enzymology, vol. 2, p. 722 (Academic Press, New York 1955).
2. BURCH, H. B.; BESSEY, O. A. and LOWRY, O. H.: Fluorimetric measurements of riboflavin and its natural derivatives in small quantities of blood serum and cells. J. biol. Chem. *175:* 457-470 (1948).

Author's address: Dr. D. GLATZLE, F. Hoffmann-La Roche & Co. Ltd., *CH-4002 Basle* (Switzerland)

7th int. Congr. clin. Chem., Geneva/Evian 1969; vol. 2: Clinical Enzymology, pp. 95–107
(Karger, Basel/München/Paris/New York 1970)

An NADH-linked Kinetic 5'Nucleotidase Assay

G. Ellis, A. Belfield and D. M. Goldberg
Department of Chemical Pathology, Royal Hospital, Sheffield

Introduction

The hydrolysis of ribonucleotide 5'-phosphates by human serum is accomplished by two enzymes, one of which, alkaline phosphatase (APase, EC 3.1.3.1., orthophosphoric monoester phosphohydrolase) is active upon a wide range of substrates; the other is the specific enzyme 5'nucleotidase (5Nase, EC 3.1.3.5., 5'-ribonucleotide phosphohydrolase). The estimation of both is of considerable diagnostic value. Until recently, the assay of 5Nase activity was complicated by a dependence upon phosphate determinations and the inadequacy of methods purporting to eliminate the contribution of APase to the hydrolysis of adenosine 5'-monophosphate (5'AMP), the substrate generally employed [6].

These difficulties were resolved by coupling liberation of adenosine to its deamination by adenosine deaminase (ADase, EC 3.5.4.4); this results in a fall in extinction at 265 nm which may be monitored continuously and permits the addition of a large excess of β-glycerophosphate. The presence of this alternative substrate suppresses hydrolysis of 5'AMP by APase [2] and we have referred to this phenomenon as 'enzyme diversion' [3]. These principles were embodied in a specific kinetic assay for 5Nase activity at 265 nm utilising 20 µl of serum [5]. Problems are encountered at this wavelength with turbid and haemolysed sera, and the high extinction of serum at 265 nm limits the amount that can be included in the reaction although an increase would be desirable where 5Nase activity is low.

These limitations do not have the same force at 340–366 nm, the wavelength used for following changes in the redox status of the pyridine nucleotides. An added attraction in operating at this wavelength is the possibility of using a wide range of instruments not capable of functioning in the ultraviolet or of supplying monochromatic light, as well as several that have been recently introduced for the kinetic determination of enzyme activities at this wavelength.

This has been achieved by coupling the release of ammonia by the reactions previously outlined to the reductive amination of oxoglutarate catalysed by glutamate dehydrogenase (GDH, EC 1.4.1.2). The method presented incorporates a linked reaction that has general application in the kinetic assay of ammonia-liberating enzymes.

Method

Principle

The following reactions are involved

$$5'AMP \xrightarrow[Mg^{++}]{5Nase} adenosine + Pi$$

$$Adenosine \xrightarrow{ADase} inosine + NH_4^+$$

$$2\text{-oxoglutarate} + NH_4^+ \xrightarrow{GDH} L\text{-glutamate}$$

$$NADH \rightarrow NAD^+$$

Reagents

All were analytical grade, unless otherwise stated, and dissolved in glass-distilled water. Glassware was cleaned by immersion in 50% ($^v/_v$) HNO_3.

1. Tris/HCl buffer, 0.1 M, pH 7.9 at 37° C, containing NaCl in final concentration of 0.8 M. Stable at least one month at 4° C.
2. $MgCl_2$, 100 mM, prepared by dissolving 403 mg MgO in approximately 20 ml lN HCl and diluting to 100 ml with water. Stable indefinitely at room temperature.
3. 2-oxoglutarate, 0.2 M, pH 6.9. Stable at least 3 months at -20° C.
4. NADH, disodium salt (Boehringer), 10 mg per ml in water, freshly made each day.
5. ADase (Boehringer, 2 mg per ml, in 2.8 M ammonium sulphate) dialysed against 10 mM phosphate buffer, pH 6.0, until free of ammonia, mixed with an equal volume of glycerol, and stored at -10° C. Loses activity at rate of 25% per month. (ADase in glycerol will shortly be available from Boehringer.)

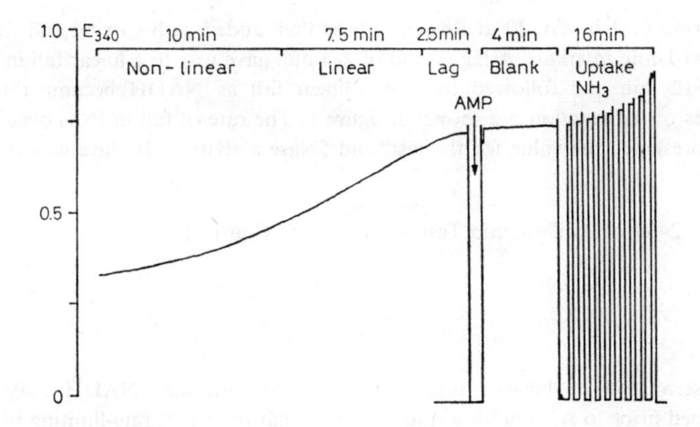

Fig. 1. Reactions in the 5Nase assay. After removal of endogenous NH_3, and measurement of its spontaneous generation (blank), AMP is added and the reaction measured over the linear portion (Unicam SP 800 Recording Spectrophotometer).

6. Glutamate dehydrogenase (Boehringer, 450 units per ml 50% glycerol). Stable at least 1 year at 4°C.
7. Sodium β-glycerophosphate 0.15 M. Stable one month at 4°C.
8. 5'AMP, disodium salt (Boehringer), 20 mM. Stable 3 months at –20°C.

Technique

The following were added to a silica cuvette with a 1-cm light-path:

1.5 ml Reagent 1
0.3 ml Reagent 2
30 μl Reagent 3
40 μl Reagent 4
30 μl Reagent 5
50 μl Reagent 6
0.6 ml Reagent 7
Water to 2.8 ml

In practice, a bulk mixture of the above reagents was prepared, 2.8 ml being taken per cuvette; no deterioration was evident over an 8-h period at room temperature. 200 μl serum was added with stirring and the cuvette placed in the SP 800 Spectrophotometer (Unicam Instruments, Cambridge, England) fitted with a constant temperature cell-holder at 37°C, SP 850 Range Expander and SP 20 External Recorder. The extinction at 340 nm was monitored at 2-min intervals. A rapid fall in extinction took place as ammonia present in the serum was consumed. After about 20 min, this fall ceased, or proceeded at a slow linear rate (< 0.003 per min) attributable to the spontaneous generation of ammonia by serum; this rate represented the value for the control and was measured over a period of 4–6 min. 30 μl Reagent 8 was then added with stirring and the reaction followed at 1-min intervals. A lag period of 2–3 min gave way to a linear fall in extinction lasting 7–12 min and followed by a non-linear fall as NADH became rate-limiting. The phases of the reaction are shown in figure 1. The rate of fall in E_{340} over this linear period represented the value for the test, and 5Nase activity in IU/litre was given by the expression:

$$2430 \times (\Delta E_{340}/\text{min Test} - \Delta E_{340}/\text{min Control})$$

Notes

1. Aged sera tend to have a high ammonia concentration. NADH may therefore be consumed prior to AMP addition to an extent that renders it rate-limiting in the linear phase of the enzyme reaction. The extinction of the serum and reaction mixture before AMP addition should be 0.9. If less, more NADH should be added.

2. The bulk reaction mixture (minus AMP) may be submitted to a series of checks. Addition of 30 μl 10 mM adenosine should cause a rapid fall in E_{340}. Addition of 30 μl Reagent 8 may initiate a fall in E_{340} which should be complete in 4 min and should not exceed 0.030; more than this indicates unacceptable contamination of 5'AMP by adenosine or ammonia, or contamination of ADase or GDH by 5Nase

Results and Discussion

Glutamate Dehydrogenase

The amount included in the assay is limited by expense, stability, and effectiveness. The more GDH is present, the shorter the lag and the more rapid is the conversion of ammonia (fig. 2), while the rate of fall in E_{340} during the linear phase of the reaction is also increased (fig. 3). In the presence of buffer, Mg^{++} ions, β-glycerophosphate and ADase, GDH is perfectly stable; but on addition of NADH, turbidity ensues leading to a steady increase in extinction at E_{340}. This can be prevented by the addition of NaCl and serum, but is also partially dependent upon the batch of GDH. With all batches tested, GDH was stable in the presence of 200 μl serum and 0.4 M NaCl. With some batches, this concentration of NaCl was adequate over the range 20–200 μl serum; with others, an increase was necessary as the amount of serum was decreased, 0.8 M NaCl being required in the absence of serum. NaCl in concentrations up to 0.4 M had little effect upon the activity of GDH, ADase or of 5Nase in the linked system. Above

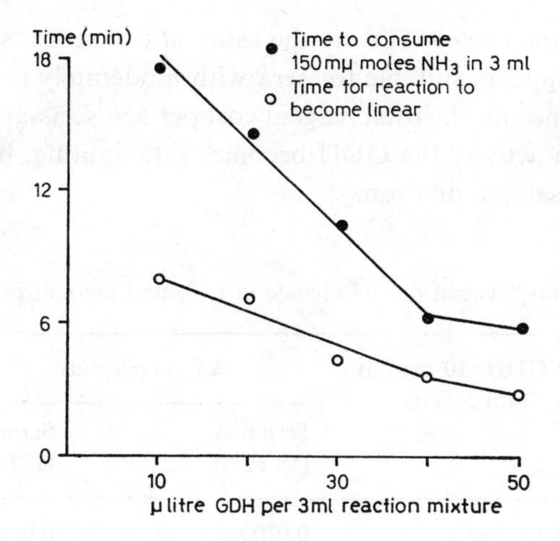

Fig. 2. Effect of increasing GDH concentration in standard assay upon rate of removal of ammonia and upon extent of the lag period.

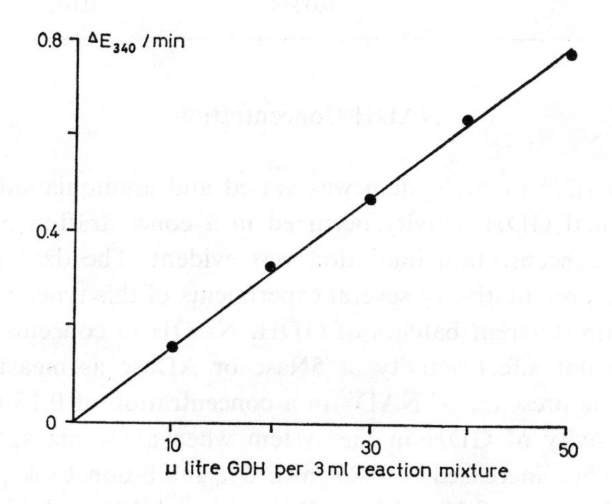

Fig. 3. Dependence of reaction rate upon GDH concentration in assay of 1 μmole NH_3.

this concentration, inhibition of the linked 5Nase reaction occurred and reached 35% at 0.8 M. A similar inhibition of ADase and GDH activity occurred in 0.8 M NaCl.

The effect of increasing GDH in the assay of two sera is shown in table I. Fifty μl GDH appears suitable for sera with moderately raised 5Nase activity; with this amount, the total reagent cost per 5Nase assay is two shillings. For sera of high activity the GDH becomes rate-limiting, but this problem is overcome by sample dilution.

Table I. Relationship between rate of change in E_{340} and amount of GDH per cuvette

μl GDH (10 mg/ml) per 3 ml cuvette	Δ E_{340} per min	
	Serum A (58 IU/l)	Serum B (177 IU/l)
10	0.0100	0.0120
20	0.0155	0.0278
30	0.0192	0.0356
40	0.0212	0.0444
50	0.0233	0.0523
100	0.0246	0.0688
150	0.0238	0.0728

NADH Concentration

When the NADH in the system was varied and ammonia substituted for serum, maximal GDH activity occurred in a concentration of 0.15 mM; beyond this concentration inhibition was evident. The data presented in figure 4 are representative of several experiments of this type, but variations were seen with different batches of GDH. NADH in concentrations up to 0.3 mM did not affect activity of 5Nase or ADase as measured by this technique. The presence of NAD^+ in a concentration of 0.15 mM did not affect the activity of GDH in the system when ammonia substituted for serum, but when increased to 0.25 mM, 5% inhibition took place. In the 5Nase-linked system, 0.15 mM NAD^+ produced 15% inhibition.

2-oxoglutarate Concentration

The concentration employed (2 mM) is 10-fold that of 5'AMP. A 10 mM concentration did not affect the activity of GDH when ammonia was

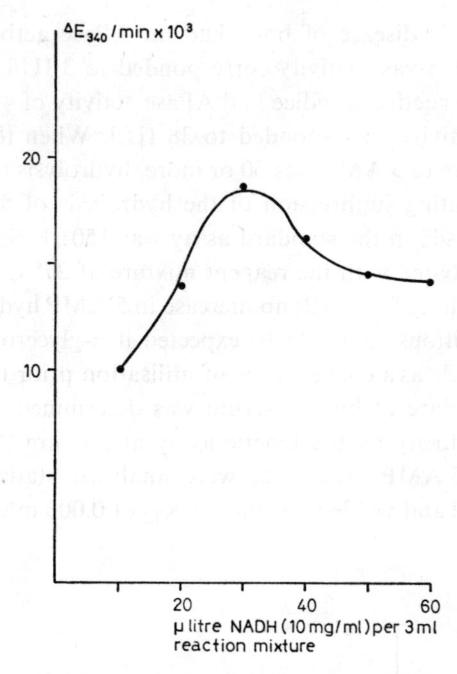

Fig. 4. Effect of increasing NADH concentration on ammonia utilization. All other components as in standard 5Nase assay except that 1 μmole NH_3 substituted for serum.

substituted for serum, or the activity of 5Nase or ADase measured by the kinetic assay at 265 nm [5].

Adenosine Deaminase

The activity of a serum with 5Nase activity of 120 IU/l was not affected when the ADase was varied between 15–200 % of the amount normally incorporated. Adequate allowance is therefore made for the slow deterioration in ADase activity already mentioned. Assuming the 5Nase activity of serum to be 15 IU/l and the other enzymes to have activity measured under optimal conditions, the ratio 5Nase: ADase:GDH is 1: 800: 7500.

Concentrations of AMP and β-glycerophosphate

Figure 5 shows the effect of varying the concentration of β-glycerophosphate with an AMP concentration maintained at 0.1 mM. The serum from a patient with Paget's disease of bone had an APase activity of 357 units/100 ml [9] and the 5Nase activity corresponded to 3 IU/l. The serum from a patient with obstructive jaundice had APase activity of 45 units/100 ml [9] and the 5Nase activity corresponded to 38 IU/l. When the molar ratio of β-glycerophosphate to 5'AMP was 50 or more, hydrolysis of 5'AMP reached a minimum indicating suppression of the hydrolysis of 5'AMP by APase. The molar ratio used in the standard assay was 150: 1. Sera of high APase activity were incubated with the reagent mixture at 37° C for periods of up to 2 h before addition of 5'AMP; no increase in 5'AMP hydrolysis took place under these conditions, as might be expected if β-glycerophosphate fell to sub-inhibitory levels as a consequence of utilisation prior to AMP addition.

The K_M for 5Nase of human serum was determined by measuring the initial reaction velocity by the kinetic assay at 265 nm [5] over the range 0.001–0.11 mM 5'AMP. The data were analysed statistically according to WILKINSON [13] and yielded a value for K_M of 0.006 mM with a Standard

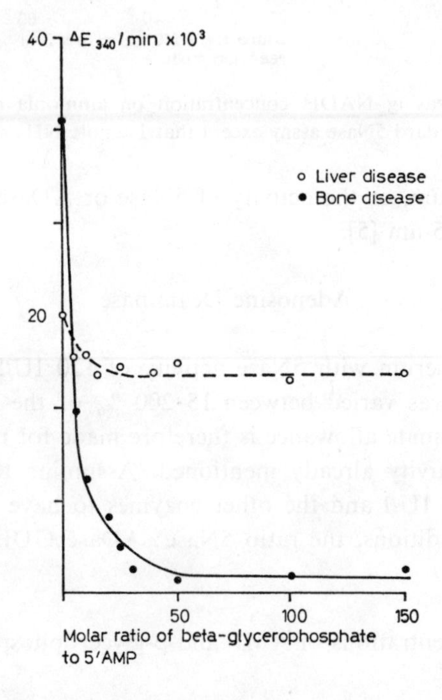

Fig. 5. Hydrolysis of 0.1 mM AMP in the presence of varying β-glycerophosphate concentration by sera from patients with liver disease and bone disease.

Error of ± 0.003 mM. The final concentration of 5'AMP in the standard procedure was 0.2 mM. This should sustain 97% of the theoretical V_{max}. The effect of varying AMP concentration while keeping the molar ratio of AMP to β-glycerophosphate constant was tested with a serum of high 5Nase and high APase activity (fig. 6). With β-glycerophosphate in 150-fold excess, a plateau was reached at 0.16 mM AMP concentration, but with 15-fold excess, activity continued to increase with increasing AMP concentration. This increase was probably due to failure of β-glycerophosphate to saturate the APase present.

Magnesium Concentration

A final concentration of 10 mM Mg^{++} ions is optimal for 5Nase activity of human serum and tissues [1, 4, 7, 10, 12]. This was the concentration used in the present method; this concentration of Mg^{++} ions had no effect on ADase but inhibited GDH by 23%.

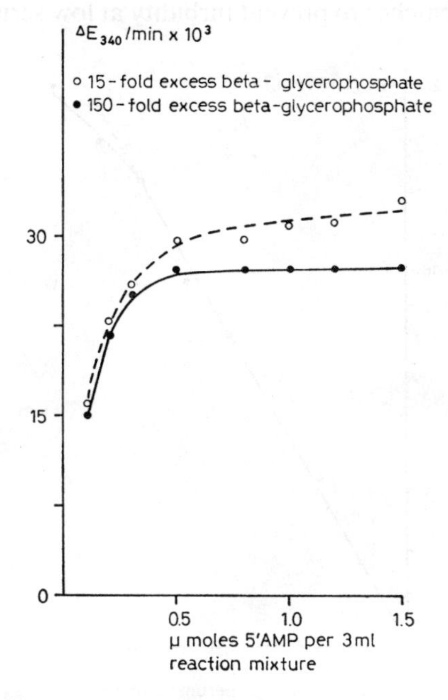

Fig. 6. Effect of increasing 5'AMP concentration, in the presence of a 15- and 150-fold excess of β-glycerophosphate, upon hydrolysis of 5'AMP by a serum with high 5Nase (65 IU/l) and high APase (102 King Units/100 ml) activities.

Buffer and pH

The pH optimum of 5Nase is 7.7–8.0 [1, 4, 8, 11] and was not re-examined with the present method. The activity in the presence of Tris buffer is 10% higher than in the same concentration of Veronal which was superior to any other buffer capable of being used in a colorimetric assay utilising the Berthelot reaction [1].

Linearity of Enzyme Assay

The rate of fall in E_{340} was proportional to the volume of serum included in the assay provided the rate did not exceed 0.020/min. It is therefore recommended that assays exceeding this rate (i.e. activity > 50 IU/l) be repeated using a dilution of the original, or less serum. Figure 7 presents results obtained using a serum with 5Nase activity of 165 IU/l and is typical of several experiments carried out with sera of different activities employing different batches of GDH. As mentioned, the concentration of NaCl had to be increased with some batches to prevent turbidity at low serum concentration.

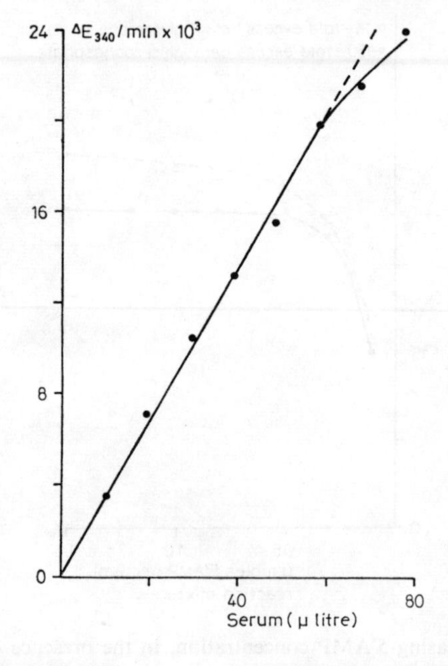

Fig. 7. Response of standard 5Nase assay to increasing amounts of serum. Projection of linearity indicated by broken line.

Precision of Enzyme Assay

Twelve replicate analyses carried out on a single serum gave a mean ΔE_{340} of 0.0240 (s = 0.0006); the coefficient of variation was thus 2.5%. Duplicate determinations were also made on two separate occasions on 10 sera with 5Nase activity ranging from 2–53 IU/l. Each determination differed from the mean of its pair by an average of 13%, the difference being greater at low activity and less at high activity; but the sum of the activities on the first set of determinations differed from the sum of the second by 3.5%.

Comparison with Kinetic Assay at 265 nm

Determinations were made with the present method and the kinetic assay at 265 nm [5] on 77 sera with 5Nase activity ranging from 1–324 IU/l. The mean activity obtained by the present method for the entire group was 65.2% of the value obtained at 265 nm, samples with activity in excess of 50 IU/l being assayed on a suitable dilution. For samples not requiring dilution, activity at 340 nm was 68.0% of that obtained at 265 nm. For the entire series the value of the correlation coefficient was 0.915 and the regression equation was

$$\text{Activity}_{340} = 0.55\ \text{Activity}_{265} + 7.82$$

The S.E. of the Slope and Intercept were 0.03 and 1.93 respectively.

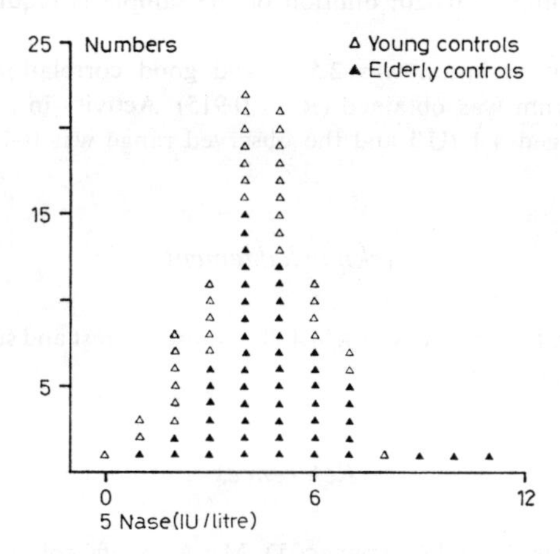

Fig. 8. Distribution of 5Nase activity in two groups of normal subjects.

Normal Range

This was established using sera from 38 healthy laboratory staff aged 18–32 (22 males; 16 females) and 50 subjects whose average age was 73 years (24 males; 26 females) attending a well-patient screening clinic. The data are presented in figure 8. The mean for the entire group was 4.4 IU/l and the observed range 0–11 IU/l. This compares with a mean and range for the assay at 265 nm of 5.2 and 0–15 respectively [5] and is in line with the lower values obtained with the present method.

Summary

A method is described for the continuous spectrophotometric determination of 5'nucleotidase (5Nase) activity at 340 nm, achieved by coupling hydrolysis of AMP to the generation of ammonia from adenosine by adenosine deaminase; this ammonia is coupled to the reductive amination of 2-oxoglutarate by glutamate dehydrogenase (GDH) and NADH. The reaction is carried out in Tris buffer, pH 7.9, in the presence of 10 mM $MgCl_2$, and β-glycerophosphate suppresses hydrolysis of AMP by alkaline phosphatase.

Optimal concentrations of link-enzymes and substrates were established, and NaCl in final concentration of 0.4 M was required to maintain GDH in solution in the presence of NADH. Pre-incubation of all the constituents prior to addition of AMP removes ammonia from serum and allows the spontaneous formation of ammonia in the system to be determined. A lag period follows addition of 5'AMP and gives way to a linear fall in E_{340} which is proportional to enzyme concentration up to a $\triangle E_{340}$ per min of 0.020; dilution of the sample is required where this rate is exceeded.

The coefficient of variation was 2.5%, and good correlation with a published kinetic assay at 265 nm was obtained (R = 0.915). Activity in a control population of 88 subjects averaged 4.4 IU/l and the observed range was 0–11 IU/l.

Acknowledgement

We are grateful to Mr. DAVID GILES, F.I.M.L.T., whose interest and suggestions prompted the present investigation.

References

1. BELFIELD, A.; ELLIS, G. and GOLDBERG, D. M.: A specific colorimetric 5'nucleotidase assay utilising the Berthelot reaction. Clin. Chem., *16*: 396-401 (1970).

2. BELFIELD, A. and GOLDBERG, D. M.: Inhibition of the nucleotidase effect of alkaline phosphatase by β-glycerophosphate. Nature, Lond. *219:* 73–75 (1968).
3. BELFIELD, A. and GOLDBERG, D. M.: Enzyme diversion applied to the kinetic estimation of glucose-6-phosphatase activity. Life Sci. *8:* 129–135 (1969).
4. BELFIELD, A. and GOLDBERG, D. M.: Activation of serum 5'nucleotidase by magnesium ions and its diagnostic applications. J. clin. Path. *22:* 144–151 (1969).
5. BELFIELD, A. and GOLDBERG, D. M.: Application of a continuous spectrophotometric assay for 5'nucleotidase activity to normal subjects and patients with liver and bone disease. Clin. Chem. *15:* 931–939 (1969).
6. BODANSKY, O. and SCHWARTZ, M. K.: 5'Nucleotidase. Advan. clin. Chem. *11:* 277–328 (1968).
7. HARDONK, M. J. and KOUDSTAAL, J.: 5'Nucleotidase. II. The significance of 5'nucleotidase in the metabolism of nucleotides studied by histochemical and biochemical methods. Histochemie *12:* 18–28 (1968).
8. HEPPEL, L. A. and HILMOE, R. J.: Purification and properties of 5'nucleotidase. J. biol. Chem. *188:* 665–676 (1951).
9. KIND, P. R. N. and KING, E. J.: Estimation of plasma phosphatase by determination of hydrolysed phenol with amino-antipyrine. J. clin. Path. *7:* 322–326 (1954).
10. PERSIJN, J. P.; SLIK, W. VAN DER; KRAMER, K. and RUIJTER, C. A. DE: A new method for the determination of serum nucleotidase. Z. klin. Chem. klin. Biochem. *6:* 442–446 (1968).
11. REIS, J. L.: The specificity of phosphomonoesterases in human tissues. Biochem. J. *48:* 548–551 (1951).
12. SCHWARTZ, M. K. and BODANSKY, O.: Properties of activity of 5'nucleotidase in human serum, and applications in diagnosis. Amer. J. clin. Path. *42:* 572–580 (1964).
13. WILKINSON, G. N.: Statistical estimations in enzyme kinetics. Biochem. J. *80:* 324–332 (1961).

Author's address: Dr. D. M. GOLDBERG, Departement of Chemical Pathology, Royal Hospital, *Sheffield 1* (England)

7th int. Congr. clin. Chem., Geneva/Evian 1969; vol. 2: Clinical Enzymology, pp. 108–112
(Karger, Basel/München/Paris/New York 1970)

A New Method for Determination of Serum Nucleotidase

J. P. PERSIJN and W. VAN DER SLIK

Department of Clinical Chemistry (Head: Dr. J. P. PERSIJN), Netherlands Cancer Institute, Amsterdam, and Department of Clinical Chemistry, State University Hospital (Head: Drs. W. VAN DER SLIK), Leiden

Methods

Reagents

The 5N-ammonia method.

1. Adenosine deaminase solution (ADA). ADA stabilised with 50% glycerol is obtainable from Boehringer Mannheim (Germany): 150 U/mg; 5 mg/ml.

2. Buffer solution. Dissolve 4.20 g sodium diethylbarbiturate (Veronal sodium) and 6.30 g $MgSO_4.7H_2O$ (A.R. grade) in about 800 ml water. Adjust the pH to 7.5 with 1N HC1 and dilute to 1000 ml with water. The solution can be used for a month if stored at 4° C.

3. Buffer/ADA solution. 0,1 ml ADA (adenosine deaminase) in glycerol 50% (150 U/mg; 5 mg/ml) is diluted with 150 ml buffer solution. Stable for 2 weeks at 4° C.

4. Buffered substrate/ADA solution. Dissolve 0.250 g adenosine-5'-monophosphate ($C_{10}H_{12}O_7N_5PNa_2 6H_2O$) in 100 ml buffer/ADA. This solution is prepared fresh before use.

5. Ethylendiaminetetra-acetic acid dipotassium salt (EDTA). Dissolve 5.6 g EDTA. 2 H_2O in water and dilute to 50 ml.

6. Phenol colour reagent (concentrated). Dissolve 50.0 g phenol (A.R. grade) and 0.25 g disodiumpenta-cyanonitrosylferrate (A.R. grade) in water and dilute to 1000 ml. Stable at least two months if kept cool and in an amber bottle protected from light.

7. Phenol colour reagent/EDTA solution. Dilute 1 volume concentrated phenol colour reagent with 4 volumes water. To 100 ml solution is added 2 ml EDTA solution. This solution must be prepared fresh before use.

8. Alkali-hypochlorite reagent (concentrated). Dissolve 25.0 g sodium hydroxide (A.R. grade) in 60 ml water. Add 72 ml of a commercial sodium hypochlorite solution containing 0.5 M NaC10 (BDH, 1 N in 0.1 N NaOH) and dilute to 1000 ml. Care should be taken that the solution is at least 0.07N to J_2 since otherwise erroneous results will be obtained. Stable at least two months if kept cool and in an amber bottle protected from light.

9. Alkali-hypochlorite reagent (diluted). Dilute 1 volume concentrated alkali-hypochlorite reagent with 4 volumes water. This solution must be prepared fresh before use.

10. Adenosine standard (3.74 μM per ml). Dissolve 100 mg adenosine (Boehringer) in saturated benzoic acid and dilute to 100 ml. The standard is stable for several months in the refrigerator.

Procedure

The detailed standard procedure is as follows: to a series of test tubes, placed in a water-bath at 37° C, different amounts are added as outlined in table I.

Table I. Standard procedure

	Buffer/ADA (ml)	Buffer/ substr./ADA (ml)	Standard (ml)	Serum (ml)
Unknown	–	1.0	–	0.1
Standard	1.0	–	0.1	–
Unknown blank	1.0	–	–	0.1
Standard blanks	1.0	–	–	–

Mix thoroughly and stopper. Incubations of the unknown is for exactly 60 min. The blanks and the standard need not be incubated for exactly 60 min. After incubation, 5 ml of diluted phenol colour reagent (containing EDTA) are added to each tube. Immediately after mixing, 5 ml of the diluted alkali-hypochlorite reagent are added. The time between these additions should be kept as short as possible. The tubes are stoppered again and left in the water-bath for a period of 30 min.

A practical point we should like to stress is that the diluted alkaline hypochlorite reagent is promptly mixed after the addition to the phenol etc. containing test-mixture. In the routine analysis the phenol and the hypochlorite reagents are added to the test mixtures by means of Cornwall continuous pipeting outfits having a two-way valve. The force by which these reagents can be injected in the test mixtures by pressing the plunger spring achieves a rapid mixing of both reagents with the contents of the test tube, which is a condition for high reproducibility of the test.

Readings of the absorbances of the unknown (A_x), standard (A_s) are made against their corresponding blanks at 625 nm using 1 cm cells.

Calculation

$$\frac{A_x}{A_s} \,.\, C_s \,.\, 16{,}7 = \text{International milli Units per ml serum.}$$

C_s = concentration standard in μM per ml.

The factor 16.7 = $0.1 \,.\, \frac{10000}{60}$ is derived from:

sample volume: 0.1 ml.

IU is expressed per liter serum: 10000.

IU is expressed per minute: 60.

Comments

1. The interference by bone alkaline phosphatase in the 5N assay can be suppressed by the addition of phenylphosphate to the incubation medium as described under reagent 4. PhP does not interfere with the colorimetric determination of adenosine and can be omitted from the blanks. For that purpose the preparation of the buffered substrate/ADA solution is changed as follows:

Dissolve 0.250 g adenosine-5-monophosphate ($C_{10}H_{12}O_7N_5PNa_26H_2O$) and 0.215 g phenyl disodium orthophosphate ($C_6H_5Na_2PO_42H_2O$) in 100 ml buffer/ADA. The solution is prepared fresh before use.

Incubation time can be extended till 75 min, in this case calculation is according to:

$$\frac{A_x}{A_s} \cdot C_s \cdot 13{,}33 = \text{mU/ml}.$$

If $C_s = 3{,}74\ \mu\text{Mol/ml}$:

$$\frac{A_x}{A_s} \cdot 50 = \text{mU/ml}.$$

The transition from normal to elevated levels of 5N in the presence of PhP corresponds to extinction values A_x (1 cm cell, 625 nm, 75 min incubation) of approximately 0.140.

2. Some commercial batches of AMP or ADA have been found to be contaminated with free adenosine and active AMP-deaminase respectively.

In such cases for each series of tests for 5N the following control must be done: a tube containing 1 ml buffer/substrate/ADA is incubated for 60 min at 37° C after which the reagents for ammonia are added as described in the standard procedure. Absorbance ($A_{CORR.}$) is read at 625 nm against a blank from which the substrate has been omitted. From all measured values of A_x this amount of $A_{CORR.}$ has to be substracted.

3. *Activity of ADA.* To a tube containing 1.0 ml buffer/ADA is added 0.2 ml adenosine stock solution. After stoppering and standing at 37° C for 2 min, the reagents for ammonia are added as described above. The absorbance is read at 625 nm against a blank which contains water instead of adenosine solution and should not be below 1.100.

The complete conversion of an amount of adenosine corresponding to a 5N activity of about 120 mU/ml within 2 min was considered to fullfill the requirements of the test.

Summary

A new method for the determination of serum nucleotidase is described; its accuracy and reproducibility have been investigated. For determinations in serum the coefficients of variation were 2.5–3.5%.

The serum 5N activity is measured by the amount of NH_3 liberated from adenosine after incubation with the enzyme adenosine deaminase (ADA). In the method reported, serum, substrate (AMP) and excess of ADA are incubated together, after which the ammonia originating from the adenosine is determined photometrically by means of the Berthelot reaction. EDTA is added to the phenol reagent to prevent precipitation of Ca and Mg salts by the alkaline hypochlorite. The present method has a much higher sensitivity than the usual methods based on inorganic phosphorus determination.

Further advantages of the method are: simple manipulations with a minimum of pipeting; deproteinization is avoided and the stock reagents are stable: ADA stabilised with glycerol can be kept for 1 year. Interference by pathological serums, introducing errors in the determination, have not been observed in recovery experiments.

References

1. PERSIJN, J. P.; SLIK, W. VAN DER; KRAMER, K. and RUYTER, C. A. DE: Z. klin. Chem. *6:* 442 (1968).
2. PERSIJN, J. P.; SLIK, W. VAN DER; TIMMER, C. J. and BON, A. W. M.: Z. klin. Chem. *7:* 199 (1969).
3. PERSIJN, J. P.; SLIK, W. VAN DER and TIMMER, C. J.: Clinical biochemistry. Canad. Soc. clin. Chemists *2:* 335 (1969).
4. PERSIJN, J. P.; SLIK, W. VAN DER and BON, A. W. M.: Z. klin. Chem. *7:* 493 (1969).
5. PERSIJN, J. P.; SLIK, W. VAN DER, TIMMER, C. J. and REIJNTJES, C. M.; submitted to Z. KLIN. Chem. (1970).
6. SLIK, W. VAN DER, PERSIJN, J. P.; ENGELSMAN, E. and RIETHORST, A.: Clinical Biochemistry, *3:* 59 (1970).

Authors' addresses: Dr. J. P. PERSIJN, Department of Clinical Chemistry, Netherlands Cancer Institute, *Amsterdam*, and Dr. W. VAN DER SLIK, Department of Clinical Chemistry, State University Hospital, *Leiden* (The Netherlands)

7th int. Congr. clin. Chem., Geneva/Evian 1969; vol. 2: Clinical Enzymology, pp. 113–120
(Karger, Basel/München/Paris/New York 1970)

Clinical Evaluation of a New Amylase Method Using Amylose Azure (Remazol Brilliant Blue Amylose) Substrate

D. A. PRAGAY and M. E. CHILCOTE

Erie County Laboratories at the Meyer Memorial Hospital and Department of Biochemistry, State University of New York at Buffalo, Buffalo, New York

In 1967, RINDERKNECHT, WILDING and HAVERBACK published a new method [8] for the assay of amylase based on the incubation of the enzyme with Remazol Brilliant Blue Amylose suspension. Enzymatic hydrolysis released the dye into the supernatant fluid. Enzymatic activity was terminated by adding acetic acid, the suspension centrifuged and the reaction quantitated by colorimetric measurement. RINDERKNECHT *et al.* studied only crystalline porcine pancreatic amylase at 37° C.

In 1968, a similar method was reported [2] with the added advantage of a soluble substrate, but this substrate has not been made available at the time of this work.

Recently another amylase method [6] was reported using chlorotriazine amylose as substrate, which is not yet available commercially at the time of this report.

The method of RINDERKNECHT *et al.* was adapted in our laboratory to assay human serum and urine amylase. Presented earlier in a preliminary form [7] this report deals with the procedural modifications and the evaluation of this method as well as the apraisal of the clinical significance.

Materials and Methods

Amylose Azure (Calbiochem)[1] powder was suspended in a phosphate buffer and mixed with serum aliquots as described below in covered 25 ml Erlenmeyer flasks. During

[1] The Remazol Brilliant Blue is a stained starch powder. It is manufactured and marketed (since January, 1969) by Calbiochem California under the commercial name Amylose Azure.

the incubation the mixture was kept in suspension by constant shaking in a controlled temperature water bath. After stopping the reaction and after centrifugation the terminal color absorbance of the supernatant was read either in a Spectronic 20 (Bausch & Lomb) or Gilford 300 Spectrophotometer at 590 nm.

The following optimal conditions were found:

pH: pH of 6.9 was found to be optimal for both phosphate and cacodylate buffer.

Temperature: 50° C was found to be optimal since higher temperatures (55–60° C) often led to stubborn cloudiness in the supernatant while the lower temperature of 37° C did not allow sufficient enzymatic activity within the desired short incubation of 15 min. The activity at different temperatures was linear up to 70 min (see fig. 1).

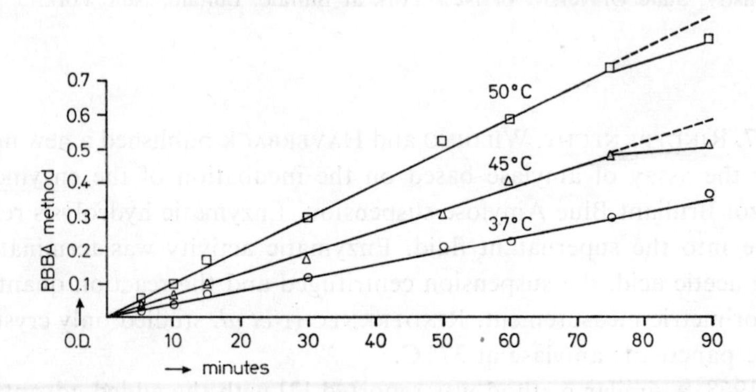

Fig. 1. Time curve of the serum amylase activity at different temperatures.

NaCl concentration: the optimal concentration varied with the buffer used, i.e., 0.1–0.3% for phosphate and 3% for cacodylate buffer.

Substrate concentration: as a compromise 2% substrate concentration was chosen for this heterogeneous system since a 1% suspension in general gave low absorbancy and economical considerations argue against higher concentrations.

Conditions for stopping the reaction: many different reagents can be used to terminate the enzymatic hydrolysis (sulfosalicylic acid, tannic acid, acetic acid, acetone, ethyl alcohol, etc., or their combination). In our hands best results were achieved with a 1:1 mixture of acetic acid (2.5%) and ethyl alcohol (40%). Use of the alcohol alone led to greater color absorbancy but cloudiness was occasionally encountered.

One has to choose the proper combination of all the above parameters in order to satisfy the requirements in the laboratory.

The following two combinations were used in our laboratory with almost equal efficiency:

Method I

4.5 ml of 2% Amylose Azure suspended in 0.05 M phosphate buffer (pH 6.9) with 3% NaCl is mixed with 0.5 ml of serum and incubated for 15 min at 50°C. The reaction is stopped by adding 2 ml of a 1:1 mixture of acetic acid (2.5%) and ethyl alcohol (40%). After centrifugation the absorbancy is read at 590 nm.

Method II

4.7 ml of 2% Amylose Azure suspended in 0.05 M phosphate buffer (pH 6.9) with 0.3% NaCl is mixed with 0.3 ml of serum and incubated for 10 min at 50° C. The reaction is stopped with 1 ml ethyl alcohol (40%). After centrifugation the absorbancy is read at 590 nm.

Although it was somewhat less efficient, one could use 0.5 ml serum, 4.5 ml buffered substrate suspension, and incubate at 37° C for 30 min.

By proportionately decreasing the amount of suspension, Method I was also adapted on a semi-micro scale by using 0.1 ml serum and the Gilford spectrophotometer.

Most of our work was done with Method I and the data presented refer to this method (although results were cross-checked with Method II).

Results

Sera with different levels of activity were assayed by using the Gomori iodometric [3], the Somogyi saccharogenic [4], and the Amylose Azure (Remazol Brilliant Blue Amylose[2]) method. Acceptable agreement was found between these methods up to 250–300 Somogyi unit levels. Activity levels above 300 units showed discrepancies (fig. 2).

The parallelity up to 300 Somogyi units can be used to «translate» the absorbancy of the Amylose Azure method into Somogyi units and to control the system with sera of known activity levels in Somogyi units.

The option of expressing the enzymatic activity in absorbancy changes or other units can, of course, be chosen but might create future problems of interpretation.

Dilution of serum samples with an enzymatic activity of approximately 300 Somogyi units or below gave proportional results, and the dilution curve was linear up to 300 Somogyi unit level (fig. 3).

The normal range was determined by assaying the serum amylase activity of 50 normal individuals and was found to be 116 $\pm$ 42 Somogyi units (1 S.D.). The 2 S.D. range was 32–200 Somogyi units. No attempt was made to determine sex differences.

[2] The expressions 'Amylose Azure' method and 'Remazol Brilliant Blue Amylose (RBBA)' method are used interchangeably in the text.

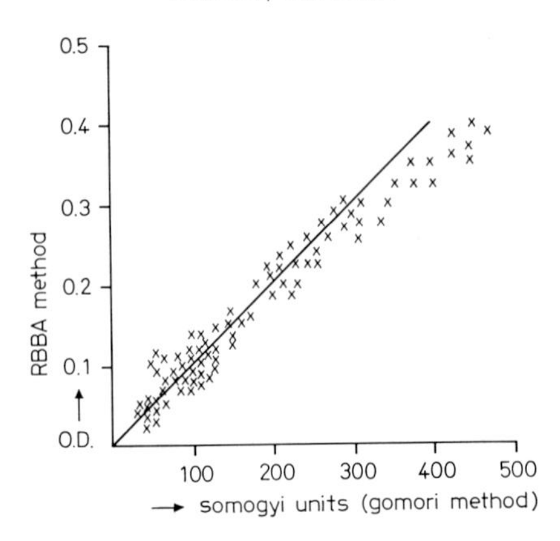

Fig. 2. Comparison of the serum amylase activity of different sera measured with RBBA method vs. Gomori iodometric method using Somogyi units.

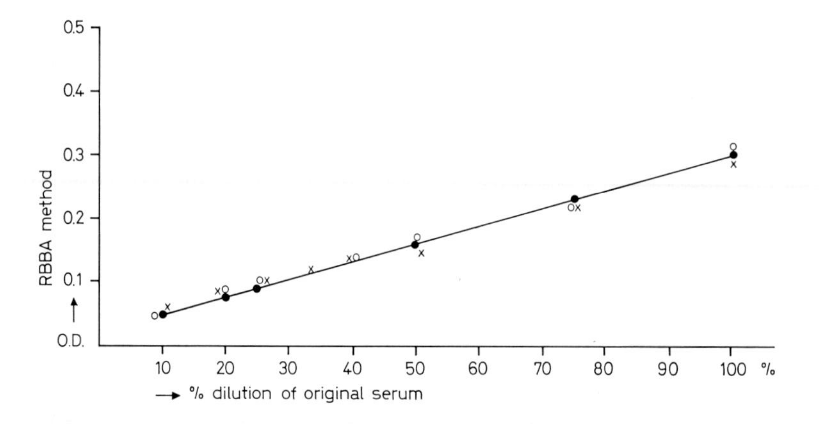

Fig. 3. Linearity of the dilution of a serum with low amylase activity (dilution was carried out with 5% albumin solution containing 0.9% NaCl). The 3 different symbols (xo.) represent the dilution of 3 different aliquots.

In order to compare the precision of different methods, pooled sera, Versatol and Hyland control sera were assayed repetitively on different days parallel with the Gomori iodometric, Somogyi saccharogenic and Remazol Brilliant Blue Amylose (RBBA) method. The precision obtained by the iodometric method showed a coefficient of variation of 8–11%, the saccharogenic method of 9–11%, and the RBBA method of 5–8% (see table I).

Table I. Comparison of the precision of three different methods

Control sera	Iodometric (Gomori) method		RBBA method		Saccharogenic (Somogyi) method	
	Activity	C.O.V.	Activity	C.O.V.	Activity	C.O.V.
Versatol E-N	172±20	11%	0.177±0.012	6%	181±20	11%
	—	—	0.214±0.019	8%	—	—
Hyland Multienzyme	270±30	11%	0.324±0.023	7%	—	—
Pooled sera	134±11	8%	0.125±0.007	5%	127±12	9%

Numbers represent amylase activities (and S. D.) in Somogyi units; C.O.V. stands for coefficient of variation. Each value represents at least 20 analyses.

In surveys obtained in different hospitals using different current amylase methods, several authors [9, 10] indicated that a coefficient of variation range of 8–13% was found, thus lending support to the better precision shown by the Amylose Azure method.

The accuracy of the Amylose Azure method was also acceptable. Sera of different patients with acute pancreatitis, hepatitis, and tuberculosis were measured with the Gomori iodometric and with the Amylose Azure method. The two methods showed parallel elevations (see table II) with the sera of these patients [1, 5][3]. Random urine samples of patients with acute pancreatitis were also analyzed in a similar way. Comparable parallel elevations were found with the two methods.

The Amylose Azure method unfortunately is not devoid of interferences. Normal and icteric sera were assayed and Hyland Pediatric Control serum with 6 mg% bilirubin, and American Monitor Bilirubin Standard (albumin dissolved) at 0.8, 2, 6, 10 mg% bilirubin levels. It was found that bilirubin interferes at abnormal range (above 3 mg%). Serious hemolysis also interferes. However, proper serum blanks can correct the results in both cases.

Lipemic sera also gives increased apparent activities but it remains very difficult to correct the results in this case.

[3] One case of a patient (L. K.) in our Hospital proved the statements of Joseph and Whitehouse [5] and of Achord [1] to be true. This statement was referring to the fact that sometimes hyperamylasemia might be due to liver diseases (hepatitis). Therefore, additional liver tests and isoamylase determinations are recommended for patients with hyperamylasemia.

Table II. A comparison of some serum amylase activities measured by both the Gomori iodometric and the RBBA methods

		RBBA method in Somogyi units	Gomori method
Patient, A. S.	5.20, 1969 ♂ 38 years		
Final diagnosis:	acute pancreatitis		
	5.20 amylase	1054	—
	5.20 SGPT	220	—
	5.28 amylase	205	215
Patient, A. C.	5.22, 1969 ♂ 82 years		
Final diagnosis:	acute pancreatitis		
	5.22 amylase	410	—
	5.23 amylase	380	440
Patient, L. K.	6.5, 1969 ♀ 46 years		
Final diagnosis:	hepatitis and cholecystitis		
	6.5 amylase	1024	1210
	6.5 SGOT	409	—
	6.5 SGPT	600	—
Patient, D. L.	7.7, 1969 ♂ 42 years		
Final diagnosis:	acute pancreatitis		
	7.8 amylase	800	950
Patient, B. S.	7.11, 1969 ♂ 35 years		
Final diagnosis:	acute pancreatitis		
	7.11 amylase	470	545
Patient, U. G.	8.7, 1969 ♀ 47 years		
Final diagnosis:	tuberculosis		
	8.8 amylase	420	480

Discussion and Conclusions

The Amylose Azure (Remazol Brilliant Blue) method presents a new approach to amylase determinations. It has certain disadvantages: the relatively high temperature, the heterogeneity of the system and the expense of the substrate (19¢ per test for the macromethod – although one should consider that the short incubation saves technician time).

On the other side this method offers considerable precision, it is quick, simple and flexible in that different parameters can be easily adjusted (temperature, time, precipitant, etc.) to the existing requirements.

The powdered substrate showed long-term stability at room temperature[4]. The buffered suspension can be stored for 2–4 weeks in the refrigerator. The colored supernatant is stable for 3 days at room temperature[5].

In conclusion, the use of a dye-bound amylose as a substrate to assay amylase activity seems to offer considerable promise. Amylose Azure (RBBA) has yielded satisfactory results in our laboratories, but it may well be replaced by newer, similar preparations now in development [2, 6].

Summary

A new amylase method – using stained starch as a substrate – was adapted for use in clinical laboratories. The optimal conditions and interferences were reported and precision and accuracy data presented as well as the normal values.

Acknowledgements

The authors wish to express their sincere thanks to the skillful assistance of Miss Geraldine Mozg and Messr. Luís Barrera and Charles Least.

References

1. Achord, J. L.: Acute pancreatitis with infectious hepatitis. J. Amer. med. Ass. *205:* 129–134 (1968).
2. Babson, A. L.; Kleinmann, N. M. and Megraw, R. E.: A new substrate for serum amylase determination. Clin. Chem. *14:* 802 (1968).
3. Gomori, G.: Assay of serum amylase with small amounts of serum. Amer. J. clin. Path. *27:* 714–16 (1957).
4. Henry, R. J.: Clinical chemistry; p. 472 (Hoeber Med. Div., Harper and Row, New York 1968).
5. Joseph, R. R. and Whitehouse, F. W.: Clin. Lab. Forum *1:* 6 (1969).
6. Klein, B.; Forman, J. A. and Searcy, R. L.: A new chromogenic substrate for the determination of serum amylase activity. Clin. Chem. *15:* 784 (1969).

[4] Under the same conditions there is little change of absorbancy from batch to batch.

[5] With precaution, the Amylose Azure powder used for blanks can be reused after drying.

7. PRAGAY, D. A. and CHILCOTE, M. E.: Clinical evaluation of an amylase method using Remazol Brilliant Blue Amylose and comparison with other amylase methods. Abstr. 7th Int. Congr. clin. Chem., Genève/Evian 1969, p. 352 (Karger, Basel/München/New York 1969).
8. RINDERKNECHT, H.; WILDING, P. and HAVERBACK, B. J.: A new method for the determination of α amylase. Experientia *23:* 805 (1967).
9. STRAUMFJORD, J. V., Jr. and COPELAND, B. E.: Clinical chemistry quality control values in thirty-three University School hospitals. Amer. J. clin. Path. *44:* 252–274 (1965).
10. TONKS, D. B.: A quality control program for quantitative clinical chemistry estimations. Canad. J. med. Tech. *30:* 38–54 (1958).

Authors' address: Dr. D. A. PRAGAY and Dr. M. E. CHILCOTE, Department of Clinical Chemistry, Erie County Laboratories at the Meyer Memorial Hospital, 462 Grider Street, *Buffalo, NY 14215* (USA)

7th int. Congr. clin. Chem., Geneva/Evian 1969; vol. 2: Clinical Enzymology, pp. 121–128
(Karger, Basel/München/Paris/New York 1970)

The Colorimetric Estimation of Arginase in Serum

W. Roman and J. Ruys
Division of Biochemistry, Institute of Medical and Veterinary Science, Adelaide

Arginase (E.C. 3.5.3.1.) is the main enzyme of the urea cycle. It catalyses the reaction: arginine $\xrightarrow[\text{arginase}]{+H_2O}$ ornithine + urea.

Accordingly it can be expected to be found in mammals mainly in the liver, and it would be useful to estimate its levels in serum in possible liver disease [2, 9, 15, 25, 27], but no real satisfactory method for the estimation of this enzyme has yet been developed. Most attempts to assay arginase in mammalian tissue were based on the measurement of urea for which there are easy methods available [1, 3, 9, 15, 17, 24, 25, 27].

There is a large excess of urea present in many homogenates and certainly in serum which makes accurate measurement of the small amounts of urea formed in the reaction very difficult. Various ways to overcome this problem have been suggested [6, 8, 16] and among them the precipitation of arginase with subsequent assaying of the precipitated protein [13].

We felt that it would be impossible to allow for any inactivation of arginase which might occur during this procedure, hence we decided rather to investigate ways of measuring arginine or ornithine. We first tried to estimate these substances spectrophotometrically, but their spectral curves are very similar and the small difference which can be picked up in the U. V. [29] is not suitable as a basis for a routine method.

Arginine forms a coloured complex with α-naphthol in the so-called Sakaguchi reaction [18–22] of which several variants have been published [4, 7, 10, 11, 23]. Ornithine will form a coloured complex with ninhydrin under acid conditions [5].

The estimation of arginine is not reliable as will be shown, but the estimation of ornithine is very reliable and a satisfactory assay method for arginase can be based on the estimation of ornithine formed as will be described in this paper.

Investigation of the Sakaguchi Reaction

Several of the methods proposed by various authors have been investigated. SAKAGUCHI [21] and later IZUMI [10] claimed that a definite compound is formed in the Sakaguchi reaction and that the various modifications proposed will eventually lead to pure Sakaguchi chromogens. WATSON [28] found that various products can be obtained with different reagents and the reaction is also not specific for arginine. We prepared a quantity of Sakaguchi chromogens with pure arginine following strictly the procedure of IZUMI [11] for the 'New Sakaguchi Reaction II' which appeared the most reliable and reproducible of all the proposed modifications. Even this, however, is not completely reproducible with pure reagents and varies slightly from batch to batch. This pure 'Arginine Sakaguchi Chromogen' was then subjected to thin film chromatography [14]. Figure 1 shows that there is not one

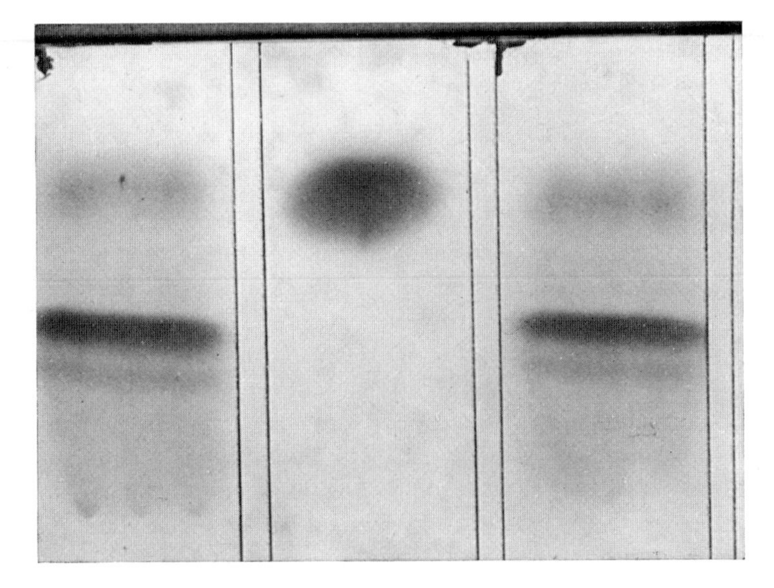

Fig. 1. Chromatographic separation of the Sakaguchi chromogen on thin layer film. Copy of a colour photograph of silica plate. The middle strip is pure arginine and the dark line on each chromatogram is arginine-ninhydrin reaction.

definite compound formed but several. It is not surprising, therefore, that one will obtain various shades according to the reagents used varying from batch to batch. Some were yellow and some orange.

It is, therefore, concluded that the Sakaguchi reaction is not a reliable method for quantitative estimation of arginine.

Investigation of the Ornithine-ninhydrin Reaction

CHINARD [5] showed that in an acid medium ornithine forms a coloured complex with ninhydrin with little or no interference from other amino acids.

Reagents

Acid reagent: 4 volumes of 6N-orthophosphoric acid and 6 volumes of glacial acetic acid. Ninhydrin reagent: 25 mg Merck ninhydrin/ml of acid reagent. Store in amber bottle. Stable for several weeks.

Procedure

1 ml of ornithine solution, 1 ml of glacial acetic acid[1] and 1 ml of ninhydrin reagent are boiled for 20 min, cooled, after which another 2 ml of glacial acetic acid were added. The optical density is read at 515 mμ.

Conditions studied

It was found that a heating of 20 min is sufficient although the original procedure suggested 60 min. The correct strength of the ninhydrin solution is important. Beer's Law is obeyed for up to 0.25 μM ornithine/ml of specimen. The ornithine standard graph is exactly reproducible; it is a straight line going through the origin.

Arginine interference was investigated; for clinical purposes arginine will not interfere significantly: 20% interference is found on 250-fold excess of arginine, as shown in table I. This small amount can be taken care of by including a serum blank.

Arginase Assay

Reagents

10 mM $MnCl_2$ solution, carbonate-bicarbonate buffer (477 mg anhydrous Na_2CO_3 + 42 mg anhydrous $NaHCO_3$ + 3 ml 1N NaOH, made up with distilled water to 100 ml). The final pH at 38° C should be 11.6. Substrate: 25 mM arginine monohydrochloride (B.D.H.), chloroform as preservative. 10% trichloroacetic acid (TCA).

[1] Some brands of glacial acetic acid, even of analytical grade, contain aldehyde which interferes with the reaction.

Table I. Interference of arginine

Content of solution/1 ml	Optical density (at 515 mμ)
Reagent blank	0
0.1 mM ornithine	27
1 mM arginine	0
5 mM arginine	1.0
10 mM arginine	2.5
25 mM arginine	5.5
50 mM arginine	13.5

Procedure

0.9 ml $MnCl_2$ solution and 0.1 ml serum are incubated in a centrifuge tube for 5 min at 37° C. Include both a test and a blank for each specimen. Add 1 ml buffer. Add 1 ml substrate to tests only. Incubate at 37° C for 30 min. Add 1 ml 10% TCA to all tubes and 1 ml substrate to the blanks. Centrifuge at 2,000 rpm for 5 min. Perform the colour reaction with 1 ml of supernatant and read O.D. at 515 mμ.

The *unit of activity* is based on the International Unit [12]: under the conditions of the assay, 1 mIU/ml of specimen is equivalent to 1.5μM of ornithine/ml.

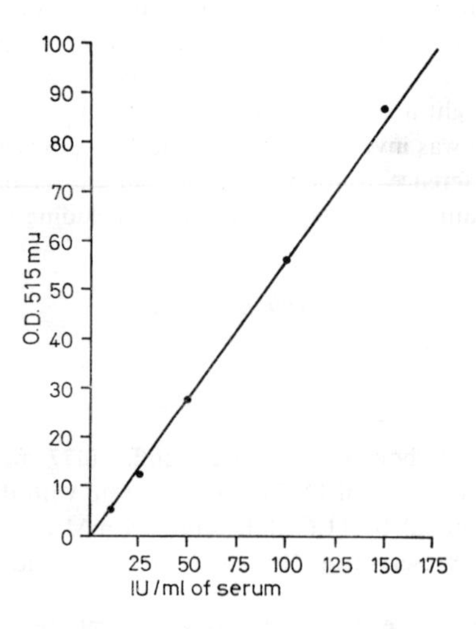

Fig. 2. Arginase standard graph.

Discussion

In developing a reliable arginase assay it was found that the estimation of urea would not be suitable as the quantity of urea present in serum is too large to make it possible to estimate accurately the small amount of urea formed by the action of the enzyme. Removal of preformed urea would lead to difficulties or partial inactivation or loss of the enzyme. The estimation of arginine by the Sakaguchi reaction is not reliable as it is not very reproducible even under extremely strict conditions as the Sakaguchi chromogen is not one single compound.

From the shape of the enzyme activity curve it is evident that an assay for enzyme activity is much more sensitive if one measures the product formed rather than the disappearance of substrate. The reaction of ornithine with ninhydrin under the conditions proposed by CHINARD [5], and modified by us, was shown to be specific for ornithine and proline only. The concentration of proline in serum is, however, very low and will not be changed by the action of arginase. The inclusion of a serum blank in which no action takes place allows to measure the specific increase in optical density due to ornithine only. The concentration of arginine is low enough not to interfere significantly with the estimation of ornithine formed.

It is, therefore, possible to base a method for the assay of arginase on the estimation of ornithine formed using the ninhydrin reaction when following strictly the conditions laid down.

Arginase has been known to need Mn^{++}ions for its activation, other ions activate to a lesser extent [3, 15, 17]. The inclusion of Mn^{++}ions in the assay proved to be rather difficult [27], as at the pH optimum of arginase, Mn tends to precipitate as $Mn(OH)_2$. A preincubation of serum with Mn^{++}ions only appeared to overcome this difficulty, as shown by the finding that no arginase activity could be measured if Mn^{++}ions are included in the reaction mixture whereas on preincubation of serum with Mn^{++}ions evidence for good arginase activity is obtained (table II).

Selection of a suitable buffer also proved to be difficult: the incubation mixture invariably reduces the pH to below optimal, the $MnCl_2$ solution being mainly responsible.

It was attempted to use the substrate arginine as a buffer, as it is known to have buffering capacity, but this proved unsuccessful. Eventually optimal pH was obtained for the incubation mixture with a carbonate buffer. The buffering capacity of the components of the incubation mixture appears satisfactory, as the pH remains stable throughout the incubation period.

Table II. Influence of Mn^{++} concentration

0.5 ml serum + 0,5 ml Mn Cl_2 solution incubated for 30 min, then 1 ml glycine buffer pH = 9.7 + 1 ml 1 mM arginine solution incubated at 37° C for 1¾ h, then
1 ml 10% TCA, centrifuged; 1 ml supernatant + 1 ml glacial acetic + 1 ml ninhydrin reagent, boil 20 min, cool + 2 ml glacial acetic, read at 515 mμ.

Serum pre-incubated with 0.5 ml	O.D.* Blank	O.D.* Test	Difference
H_2O	13.5	13.5	0
1 mM Mn^{++}	14	20.5	6.5
5	12.5	21	8.5
10	14	22.5	8.5
50	14	19	5
100	13.5	15.5	2
500	13.5	13.5	0

* Optical density.

It is possible to use as little as 0.1 ml of serum as the method is sensitive enough to detect as little as 0.1 units of arginase in this amount of serum, as 1 mM unit per ml can be estimated by this method.

Summary

A satisfactory routine method for the assay of arginase in serum has been developed. This is based on the estimation of ornithine formed. Arginase requires Mn^{++}ions for optimum activity and the optimal pH = 11.6. As little as 1 mIU/1 ml of serum can be estimated.

Acknowledgements

Some of the preliminary work on the investigation of the Sakaguchi reaction was carried out by Miss H. ROBERTSON and Miss A. YOONG. This work was supported by a financial grant from the Ames Company.

References

1. Basilio, C.; Prajoux, V. y Cabello, J.: Arginasa de la sangre y del higado en ratas intoxicadas con tetracloruro de carbone. Acta physiol. latin-amer. *10:* 159 (1960).
2. Brown, H.; Brown, M. E.; Michelson, P. and McDermott, W. V.: Urea-cycle enzymes in liver disease. J. Amer. med. Ass. *199:* 35 (1967).
3. Cabello, J.; Basilio, C. and Prajoux, V.: Kinetic properties of erythrocyte and liver arginase. Biochim. biophys. Acta *48:* 148 (1961).
4. Ceriotti, G. and Spandrio, L.: An improved method for the microdetermination of arginine by use of 8-hydroxyquinoline. Biochem. J. *66:* 603 (1957).
5. Chinard, F. P.: Photometric estimation of proline and ornithine. J. biol. Chem. *199:* 91 (1952).
6. Cornelius, Ch. E. and Freedland, R. A.: The determination of arginase activity in serum by means of gel filtration. Cornell Vet. *52:* 344 (1962).
7. Gilboe, D. D. and Williams, J. N.: Colorimetric method for rapid determination of arginase activity. Proc. Soc. exp. Biol. Med. *91:* 537 (1956).
8. Hagan, J. J. and Dallam, R. D.: Measurement of arginase activity. Analyt. Biochem. *22:* 518 (1968).
9. Illiano, G.; Soscia, M.; Picciafuoco, P.; Giordano, F., e Pietra, C. Della: Azione inibitrice del fegato di pulcino sull'attività arginasica di fegato di ratto. Biochim. appl. *13:* 107 (1966).
10. Izumi, Y.: New Sakaguchi Reaction. Analyt. Biochem. *10:* 218 (1965).
11. Izumi, Y.: New Sakaguchi Reaction II. Analyt. Biochem. *12:* 1 (1965).
12. King, E. J. and Campbell, D. M.: International enzyme units: an attempt at international agreement. Proc. 4th Int. Congr. clin. Chem., Edinburgh 1960, p. 185. (Livingstone, Edinburgh/London 1961).
13. Manning, R. T. and Grisollia, S.: Serum arginase activity. Proc. Soc. exp. Biol. Med. *95:* 225 (1957).
14. Pataki, G.: Dünnschicht Chromatographie in der Aminosäure- und Peptid-Chemie, p. 79. (de Gruyter, Berlin 1969).
15. Pelikan, V.; Kalab, M. and Tichy, J.: Determination of arginase activity in blood in epidemic hepatitis. Clin. chim. Acta *9:* 142 (1964).
16. Reynolds, J.; Folette, J. H. and Valentine, W. N.: The arginase activity of erythrocytes and leucocytes with particular reference to pernicious anemia and thalassemia major. J. Lab. clin. Med. *50:* 78 (1957).
17. Roberts, E.: Estimation of arginase activity in homogenates. J. biol. Chem. *176:* 213 (1948).
18. Sakaguchi, S.: Über eine neue Farbenreaktion von Protein und Arginin. J. Biochem., Tokyo *5:* 25 (1925).
19. Sakaguchi, S.: Über die Bindungsweise und quantitative Bestimmung des Arginins im Proteinmolekule. J. Biochem., Tokyo *5:* 133 (1925).
20. Sakaguchi, S.: A new method for the colorimetric determination of arginine. Jap. med. J. *1:* 278 (1948).
21. Sakaguchi, S.: A new method for the colorimetric determination of arginine. J. Biochem., Tokyo *37:* 231 (1950).

22. SAKAGUCHI, S.: Note on the colorimetric determination of arginine. J. Biochem., Tokyo *38:* 91 (1951).
23. SATAKE, K. and LUCK, J. M.: The spectrophotometric determination of arginine by Sakaguchi reaction. Bull. Soc. Chim. biol. *40:* 1743 (1958).
24. SATOH, P. S. and ITO, Y.: Quantitative rapid microassay of arginase. Analyt. Biochem. *23:* 219 (1968).
25. UGARTE, G.; PINO, M. E. and VALENZUELA, M. D.: Liver arginase activity in patients with liver cirrhosis and in patients in endogenous hepatic coma. J. Lab. clin. Med. *57:* 359 (1961).
26. VAN SLYKE, D. D. and ARCHIBALD, R. M.: Gasometric and photometric measurement of arginase activity. J. biol. Chem. *165:* 293 (1946).
27. VINCENT, D. et SEGONZAC, G.: Le dosage de l'arginase sérique en pratique clinique. Rev. Franç. Et. clin. biol. *5:* 390 (1960).
28. WATSON, D.: Chromogenicity of Sakaguchi-reactive compounds, monosubstituted guanidines and histidine. Nature, Lond. *211:* 411 (1966).
29. WARD, R. L. and SRERE, P. A.: A new spectrophotometric arginase assay. Analyt. Biochem. *18:* 102 (1967).

Authors' addresses: Dr. W. ROMAN, Director of Laboratories, Queen Victoria Hospital, *Rose Park*, (South Australia); Mr. J. RUYS, Head of Toxicology Section, Riker Laboratories (Australia Pty. Ltd.), *Thornleigh, New South Wales* (Australia)

7th int. Congr. clin. Chem., Geneva/Evian 1969; vol. 2: Clinical Enzymology, pp. 129–132 (Karger, Basel/München/Paris/New York 1970)

Properties of Glutathione Reductase (E.C.1.6.4.2.) from Normal Erythrocytes and from Erythrocytes of a Patient with a Glutathione Reductase Deficiency

G. E. J. Staal, P. W. Helleman, J. de Wael and C. Veeger

Haematological and Biochemical Department, State University Hospital, Utrecht, and Department of Biochemistry, Agricultural University, Wageningen

Glutathione reductase catalyzes the reduction of oxidized glutathione (GSSG) to reduced glutathione (GSH); the hydrogen donor is NADPH or NADH, though the latter is less effective. Both Scott [1] and Icen [2] proved that the enzyme from human erythrocytes is a flavoprotein with flavin adenine dinucleotide (FAD) as prosthetic group.

In order to explain the decreased glutathione reductase activity in the erythrocytes of a patient with a hemolytic anemia, the normal enzyme as well as the enzyme of the patient have been purified and characterized with respect to chemical and enzymatic properties [for the methods see refs. 3, 4 and 5].

The normal human erythrocyte enzyme was purified 47,000 times by column chromatography (spec. act. 165 μM NADPH oxidized min^{-1}, mg $protein^{-1}$). From the minimum MW of 56,000 per mole FAD in combination with the MW of 115,000 of the enzyme, it can be concluded that the native enzyme contains two molecules of FAD and is built up by two polypeptide chains, in agreement with previous conclusions for the yeast enzyme [6, 7]. FAD can be detached from the protein to produce the apoenzyme. Incubation of the apoenzyme with flavin mononucleotide (FMN) or with riboflavin does not restore the enzymatic activity, only FAD is active in this respect. Although FMN is not capable of restoring the catalytic activity upon incubation with apoenzyme, the protein has affinity for this flavin compound. Preincubation with low concentrations of FMN retards the restoration by FAD but ultimately the same activity level is reached. Adding high concentration of FMN after the full restoration of activity by FAD, leads to a decline of the activity (see fig. 1). If the same amount of FMN is added to the holoenzyme itself, no loss of activity could be demonstrated.

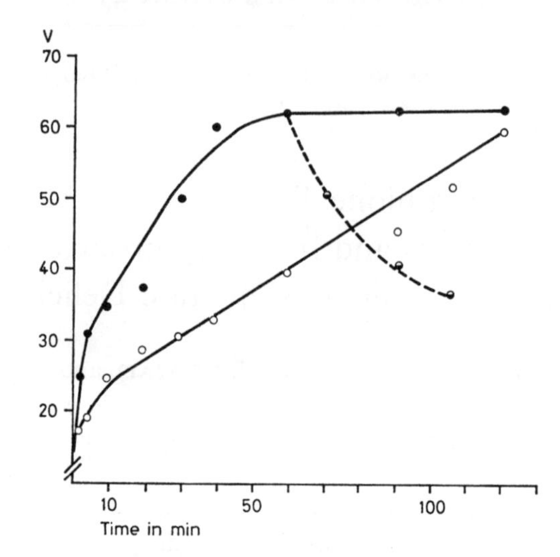

Fig. 1. Influence of FMN on the restoration of GSSG activity. 8μM FAD was added to the apoenzyme (0.05 mg/ml) at 0°C.

- ●—● preincubation at 0°C during 30 min with 5×10^{-5} M FMN
- ○—○ control, not preincubated with FMN
- ○- - -○ 5×10^{-4} M FMN added after full restoration of activity by FAD

Samples were withdrawn at different times and the activities measured at 25°C. Activities are expressed as μM NADPH oxidized/min/mg protein. Buffer 0.2 M sodium phosphate pH 6.8.

The enzyme of the patient was purified 10,000 times (spec. act. 20 μM NADPH oxidized, min^{-1}, mg $protein^{-1}$). Assuming an identical activity of the pure enzymes, this preparation is about 10% pure. The K_M values for GSSG and NADPH and the optimum pH are pratically the same as those of the normal enzyme. This is in contrast with the results obtained by WALLER [8] with enzyme from deficient patients. The activity of the patients' enzyme in either hemolysate or at the purified state can be increased when the enzyme is preincubated with FAD in the reaction mixture prior to the assay and the reaction is started by the addition of NADPH.

The enzyme needs FAD for stabilization. Heating of the normal enzyme during 1 h at 60°C in 0.1 M sodium phosphate buffer pH 7.2 does not lead to loss of activity; however, when the patients' enzyme was heated during 30 min at 55°C, about 30% of the activity was lost; on the other hand, in the presence of 10μM FAD no loss of activity could be detected. From the recombination of the abnormal apoenzyme with FAD it could be calculated that the K_{ass} for FAD was about 6 times lower than that of the normal apoenzyme. When the holoenzyme of the patient was incubated during 12 h in the presence of 10μM FMN, there was about 30% loss of activity while enzyme incubated without added FMN showed no decreased activity. The normal holoenzyme shows under identical conditions no diminished activity after incubation with FMN. The inactivation to some extent by FMN and the need for FAD during purification and for stabilization, allows the conclusion that the flavin moiety of the abnormal holoenzyme is much less tightly bound than in the normal holoenzyme. The diminished K_{ass} value for FAD binding to the apoprotein supports this conclusion. About the nature of the difference in affinity of the two proteins for FAD one can only speculate. Table I summarizes our results.

In patients with clinical symptoms of riboflavin deficiency (stomatitis for example) GLATZLE *et al.* [9] observed an activation of glutathione reductase of up to 20% and more when FAD was added to the hemolysate. Although the effect of FAD administration in the cases of GLATZLE *et al.* [9] is the same as we found in our case, we believe that the stimulating effect of FAD is caused by the diminished K_{ass} value for the abnormal enzyme. Furthermore, we found no signs of riboflavin deficiency (stomatitis, glossitis for example).

In view of the stimuling effect of FAD on the abnormal enzyme we administered a flavin preparation as therapeutic agent to the patient (10 mg FMN per day); we chose FMN because this nucleotide is *in vivo* converted into FAD. After a few weeks the enzyme activity had increased to an almost

Table I. Some properties of normal and an abnormal glutathione reductase from human erythrocytes

	Normal	Abnormal
K_M GSSG	125μM	82 μM
K_M NADPH	13.3 μM	8 μM
pH optimum	6.8	6.8
K_{ass} FAD	2×10^6 1.M^{-1} No influence of FAD and FMN on activity of holoenzyme	3×10^5 1.M^{-1} FAD increases and FMN decreases the activity of holoenzyme

K_M values are determined in 0.3 M sodium phosphate.

normal level. When this treatment was stopped the glutathione reductase activity started to decline again. The effect of administration of FAD was the same. Furthermore, during the administration of both nucleotides no increase in activity *in vitro* was found after addition of FAD to the assay. On the other hand, after incubation of the hemolysate during 12 h with an excess of FAD again a stimulation of the activity was observed. Studies are in progress to explain these results.

In view of our results, it is recommendable to add FAD to the assay mixture when a glutathione reductase deficiency exists.

References

1. SCOTT, E.; DUNCAN, I. W. and EKSTRAND, V.: Purification and properties of glutathione reductase of human erythrocytes. J. biol. Chem. *238:* 3928–3933 (1963).
2. ICEN, A.: Glutathione reductase of human erythrocytes. Scand. J. clin. Lab. Invest. *20:* Suppl. 96 (1967).
3. STAAL, G. E. J.; VISSER, J. and VEEGER, C.: Purification and properties of glutathione reductase of human erythrocytes. Biochim. biophys. Acta *185:* 39–48 (1969).
4. STAAL, G. E. J. and VEEGER, C.: The reaction mechanism of glutathione reductase from human erythrocytes. Biochim. biophys. Acta. *185:* 49–62 (1969).
5. STAAL, G. E. J.; HELLEMAN, P. W.; WAEL, J. DE and VEEGER, C.: Purification and properties of an abnormal glutathione reductase from human erythrocytes. Biochym. biophys. Acta *185:* 63–69 (1969).
6. MASSEY, V. M. and WILLIAMS, C. H.: On the reaction mechanism of yeast glutathione reductase. J. biol. Chem. *240:* 4470–4480 (1965).
7. MAVIS, R.D. and STELLWAGEN, E.: Purification and subunit structure of glutathione reductase from Bakers' yeast. J. biol. Chem. *234:* 809–814 (1968).
8. WALLER, H. D.: Reinigung der Erythrocyten-Glutathionereductase von Gesunden und Patienten mit Glutathionereductase-Mangel. Klin. Wschr. *45:* 827–832 (1967).
9. GLATZLE, D.; WEBER, F. and WISS, O.: Enzymatic test for the detection of a riboflavin deficiency. NADPH dependent glutathione reductase of red blood cells and its activation by FAD in vitro. Experientia *24:* 1122 (1968).

Authors' addresses: G. E. J. STAAL, P. W. HELLEMAN, J. DE WAEL and C. VEEGER, Haematological and Biochemical Department, State University Hospital, *Utrecht*, and Department of Biochemistry, Agricultural University, *Wageningen* (The Netherlands)

7th int. Congr. clin. Chem., Geneva/Evian 1969; vol. 2: Clinical Enzymology, pp. 133–138
(Karger, Basel/München/Paris/New York 1970)

Anémies hémolytiques congénitales avec inclusions intra-érythrocytaires et ATP bas
Etude de deux familles

P. CARTIER, A. NAJMAN, P. KAMOUN et J. P. LEROUX
Enzymologie Médicale, I.N.S.E.R.M., et Service d'hématologie, Centre hospitalier universitaire Saint-Antoine (Professeur R. ANDRÉ), Paris

Introduction

Depuis l'observation de GRIMES *et al.* [7], on sait que certaines anémies hémolytiques congénitales, caractérisées par l'apparition de corps de Heinz après splénectomie et s'accompagnant de mésobilifuchsinurie (« urines noires »), présentent des modifications biochimiques importantes:
hémoglobines instables [6, 12, 13];
modifications glycolytiques associant une glycolyse accélérée à un taux abaissé d'ATP intraglobulaire [7, 10].

La relation entre l'hémolyse, l'instabilité de l'hémoglobine et les anomalies métaboliques a été discutée par JACOB *et al.* [8], et sera évoquée ici à propos de l'étude de deux nouvelles observations familiales.

Méthodes expérimentales

Le dosage des enzymes glycolytiques [2], des intermédiaires de la glycolyse [3, 5], des activités purine-phosphoribosyltransférasiques [1] et la mesure de l'activité glycolytique globale [4], ont été précédemment décrits.

La cinétique de perméabilité aux ions sodium a été réalisée selon la technique suivante: les globules rouges lavés par une solution isotonique de $MgCl_2$ sont incubés à 37°C pendant 2 h dans un tampon isotonique pH 7,35 (phosphate mono/disodique 40 mM; KCl, 4mM; $MgCl_2$, 5mM; glucose, 20 mM; albumine bovine dialysée 25 g par litre; Na Cl, 77mM, additionné d'une dose traceuse de ^{24}Na.)

Les globules rouges sont alors lavés trois fois à 2°C par la solution isotonique de $MgCl_2$ et incubés à nouveau dans les mêmes conditions de température, dans le milieu précédent non marqué. La radioactivité du surnageant est déterminée au compteur γ après centrifugation à 2°C des prélèvements.

L'incorporation de l'adénine dans les nucléotides puriques a été effectuée sur les globules lavés au sérum physiologique puis incubés à 37° C pendant 90 min dans un milieu isotonique de pH 7,4 (NaCl, 42 mM; phosphate mono/dipotassique, 31 mM; glucose, 7,5 mM; adénine, 0,22 mM, marquée par 8–^{14}C–adénine). Après lavage au sérum physiologique à 2° C, le culot globulaire est déféqué par l'acide perchlorique, et la répartition de ^{14}C-adénine est déterminée par comptage au Tricarb-Packard sur les taches obtenues par chromatographie du surnageant sur papier DEAE-cellulose Whatman, en solvant Paladini-Leloir [11].

Résultats et discussion

Le tableau I fait apparaître l'intensité très variable du processus hémolytique d'un malade à l'autre: certains sujets bien compensés ne traduisent leur hémolyse que par une réticulocytose discrète associée à une élévation des activités enzymatiques (tableau II). Il n'existe aucune enzymopénie: toutes les enzymes glycolytiques ont été mesurées mais seuls sont présentés les résultats de l'hexokinase et de la pyruvate kinase, dont l'élévation témoigne du renouvellement globulaire rapide.

De même les résultats des dosages des substrats intermédiaires de la glycolyse, qui restent dans les limites normales, ne sont pas figurés. Les concentrations du 2,3-diphosphoglycérate éliminent un déficit en diphosphoglycérate-mutase évoqué dans une observation précédente [9].

Tableau I. Données cliniques et hématologiques

	Famille C - R			Famille L - C		
	Propositus	Fils	Frère	Propositus	Fille	Cousine germaine
Sexe	F	M	M	M	F	F
Globules rouges/mm³	3 400 000	5 350 000	4 850 000	3 700 000	3 400 000	5 100 000
Réticulocytes/mm³	357 000	294 000	669 000	172 500	333 000	127 000
Bilirubine libre (mg/1)	15	7	26	14		4
Inclusions intra-érythrocytaires	+	0	0	+	0	0
Urines	Noires	Noires	Noires	Foncées	Foncées	Foncées
Splénectomie	+	0	0	+	0	0

Tableau II

	Famille C - R			Famille L - C				
	Propositus	Fils	Frère	Propositus	Fille	Cousine germaine	Normaux	± σ
Substrats								
2,3-DPG	4170	5260	4030	4860	5300	4640	4580	610
ATP	970	1240	990	870	1060	1220	1527	180
ADP	196	143	118	170	173	158	170	21
AMP	34	45	16	33	32	54	20	
Somme adénine nucléotides	1200	1428	1124	1073	1265	1432	1717	
Enzymes								
Hexokinase	0,80	0,37	0,57	0,57	0,40	0,41	0,143	0,032
Pyruvate-kinase	6,5	5,7	5,3	7,6	6,4	7,0	2,43	0,45
Glycolyse								
Glucose consommé	4,30	2,70	3,70	3,40	1,80	1,80		
Lactate formé	3,45	2,50	2,80	2,85	1,60	1,90		
Adénine ^{14}C								
Incorporation dans	1,36		1,63	1,86				
Adénine nucléotides	1,50							

Substrats: en mμm/ml de globules rouges.
Enzymes: μm de substrat consommé/min/ml de globules rouges.
Glycolyse: 3 heures à pH 7,4 et 37° C. Valeurs malades/valeurs témoins.
Incorporation d'adénine ^{14}C dans les purine-nucléotides: pourcentage d'incorporation en 90 min, valeurs malades/valeurs témoins.

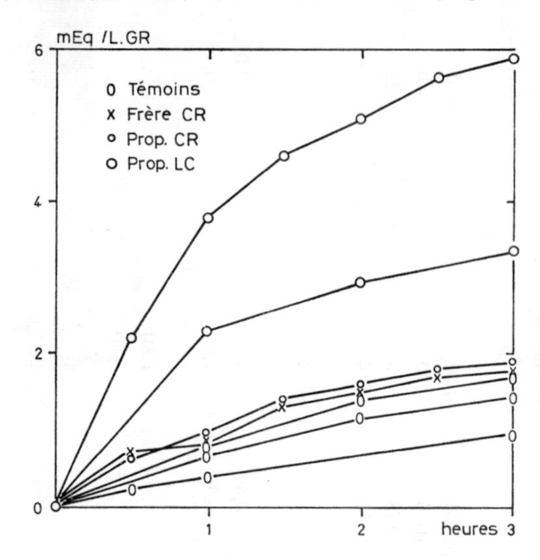

Fig. 1. Efflux du sodium à partir de globules rouges préincubés 2 h en présence de ^{24}Na.

La diminution du taux d'ATP et l'augmentation de l'activité glycolytique ont été constatées chez les six sujets: elles sont particulièrement importantes chez les trois malades cliniquement les plus atteints (les deux propositi et le frère C - R).

Cette chute du taux d'ATP n'est pas imputable à un *défaut de recharge* puisque la glycolyse est au contraire augmentée. La diminution de la somme des nucléotides adényliques pourrait évoquer un *défaut de synthèse* de l'AMP à partir des précurseurs: mais l'activité adénine-pyrophosphoribosyltransférasique est normale et l'incorporation de l'adénine-^{14}C dans les nucléotides puriques est notablement accélérée. Le niveau bas de l'ATP apparaît plus vraisemblablement lié à un *excès d'utilisation* qu'il est logique de rapporter au transport actif des ions Na^+ et K^+. Bien que la concentration érythrocytaire de ces deux cations soit normale chez les six sujets, nous avons étudié la vitesse de renouvellement des ions sodium. La figure 1, qui représente la cinétique de sortie, montre une différence importante entre les deux groupes: le propositus L - C présente une exclusion fortement accélérée, alors que dans la famille C - R les valeurs se situent à la limite supérieure de la normale. Il convient de souligner que le propositus de cette dernière famille présente une hémoglobine instable en cours d'identification[1], tandis que l'hémoglobine du propositus L - C n'obéit à aucun des critères d'instabilité actuellement connus.

Summary

Six cases from two families presenting Heinz-body hemolytic anemia have been studied. In all cases a decreased red blood cell ATP was observed: it is probably due to an excessive utilization of the nucleotide, the increased glycolytic activity being aimed at the correction of this loss. In the first family (three cases), hemoglobin appeared normal by the present tests, and the increased utilization of ATP is linked to an increased permeability of cell membranes to cations. In the second family (three cases), membrane permeability was normal, but hemoglobin, unstable: the relationship between this unstable hemoglobin and decreased ATP has not yet received a satisfactory explanation.

Bibliographie

1. Cartier, P. et Hamet, M.: Les activités purine-phosphoribosyl transférasiques des globules rouges humains. Technique de dosage. Clin. chim. Acta. *20:* 205 (1968).
2. Cartier, P.; Leroux, J.-P. et Marchand, J.-Cl.: Techniques de dosage des enzymes glycolytiques tissulaires. Ann. Biol. Clin. *25:* 109 (1967).
3. Cartier, P.; Leroux, J.-P. et Temkine, H.: Techniques de dosage des intermédiaires de la glycolyse dans les tissus. Ann. Biol. clin. *25:* 791 (1967).
4. Cartier, P.; Najman, A.; Leroux, J.-P. et Temkine, H.: Les anomalies de la glycolyse au cours de l'anémie hémolytique par déficit du globule rouge en pyruvate-kinase. Clin. chim. Acta *22:* 165 (1968).
5. Cartier, P. et Temkine, H.: Le 2,3-diphosphoglycérate et le glucose-1,6-diphosphate du globule rouge: techniques de dosage. Ann. Biol. clin. *25:* 1119 (1967).
6. Fairbanks, V. F.; Opfell, R. W. and Burgert, E. O.: Three families with unstable hemoglobinopathies (Köln, Olmsted and Santa Ana) causing hemolytic anemia with inclusion bodies and pigmenturia. Amer. J. Med. *46:* 344 (1969).
7. Grimes, A. J.; Meisler, A. and Dacie, J. V.: Congenital Heinz-body anaemia. Further evidence on the cause of Heinz-body production in red cells. Brit. J. Haemat. *10:* 281 (1964).
8. Jacob, H. S.; Brain, M. C. and Dacie, J. V.: Altered sulfhydryl reactivity of hemoglobins and red blood cell membranes in congenital Heinz-body hemolytic anemia: J. clin. Invest. *47:* 2664 (1968).
9. Löhr, G. W. und Waller, H. D.: Zur Biochemie einiger angeborener hämolytischer Anämien. Folia haemat. *8:* 377 (1963).
10. Mills, G. C.; Levin, W. C. and Alperin, J. B.: Hemolytic anemia associated with low erythrocyte ATP. Blood *32:* 15 (1968).

[1] Institut de Pathologie Moléculaire, CHU Cochin, Paris.

11. PALADINI, A. C. and LELOIR, L. F.: Studies on uridine-diphosphate-glucose. Biochem. J. *51:* 426 (1952).
12. PERUTZ, M. F. and LEHMANN, H.: Molecular pathology of human haemoglobin. Nature, Lond. *219:* 902 (1968).
13. SATHIAPALAN, R. and ROBINSON, M. G.: Hereditary haemolytic anaemia due to abnormal haemoglobin (haemoglobin Kings County). Brit. J. Haemat. *85:* 579 (1968).

Adresses des auteurs: Prof. P. CARTIER, Dr. A. NAJMAN, Dr. P. KAMOUN et Dr. J. P. LEROUX, Enzymologie Médicale, I.N.S.E.R.M., Unité 75, 156, rue de Vaugirard, *75 - Paris 15*e; Service d'hématologie, CHU Saint-Antoine, *Paris* (France)

7th int. Congr. clin. Chem., Geneva/Evian 1969; vol. 2: Clinical Enzymology, pp. 139–143
(Karger, Basel/München/Paris/New York 1970)

Etude biochimique d'une anémie hémolytique avec déficit familial en phosphohexo-isomérase

P. CARTIER, H. TEMKINE et C. GRISCELLI

Enzymologie Médicale, I.N.S.E.R.M., et Clinique Médicale Infantile (Professeur P. MOZZICONACCI), Paris

L'Ecole de VALENTINE a décrit récemment [1] une nouvelle forme d'anémie hémolytique congénitale, transmise sur le mode autosome récessif s'accompagnant d'un déficit érythrocytaire en phosphohexo-isomérase (PHI).

Peu après, nous avons étudié un second cas dont nous rapportons ici l'étude métabolique:

Nadia AM..., agée de 5 ans, présente depuis la naissance une anémie hémolytique grave ayant nécessité 14 séjours à l'hôpital et de nombreuses transfusions. Une première exploration enzymatique de ses érythrocytes (en 1966) s'était révelée négative. En fin 1968, l'enfant présentait une anémie importante ($3{,}2.10^6$ globules rouges/mm^3 avec 25% de réticulocytes. La dernière transfusion remontait à un mois et demi.

L'exploration du métabolisme glucidique (glycolyse, enzymes et intermédiaires et pyridine nucléotides) a été réalisée selon les techniques décrites antérieurement [2, 3, 4, 5].

1. L'activité glycolytique est très augmentée (3,25 μmoles de glucose consommé par ml de globules rouges soit 2,4 fois la normale), avec un rapport lactate/2 glucose normal (0, 91).

2. En rapport avec l'importante réticulocytose, les activités des enzymes glycolytiques sont augmentées et les variations sont conformes au profil glycolytique des cellules jeunes [MARCHAND et GARREAU, 6]. Seule l'activité de la phosphohexo-isomérase (PHI) est nettement abaissée de 60% (tableau I). Des variations analogues se retrouvent dans les leucocytes.

3. Le taux des intermédiaires de la glycolyse a été comparé à celui de sujets normaux et de sujets présentant une anémie auto-immune sans enzymopathie avec une réticulocytose de 20% environ (tableau II): les augmentations constatées correspondent ici aussi à la présence de réticulo-

Tableau I. Activité des enzymes glycolytiques

		Globules rouges		Leucocytes	
		Normaux	Nadia Am..	Normaux	Nadia Am..
PHM	Phosphohexomutase	0,30	0,60		
HK	Hexokinase	0,14	0,69	58	60,8
PHI	Phosphohexo-isomérase	7,24	2,90	3167	1574
PFK	Phosphofructokinase	1,30	1,51		
ALDO	Aldolase	0,38	0,67		
PTI	Triosephosphate isomérase	258	255		
GAPD	Phosphoglyceraldehyde deshydrogénase	13,2	17,6		
PGK	Phosphoglycérate kinase	28,1	25,2		
PGM	Phosphoglycérate mutase	3,5	5,0		
ENO	Enolase	1,7	3,0		
PK	Pyruvate-kinase	2,7	5,9	1373	1459
LDH	Lactico-deshydrogénase	26,1	25,8	1098	1081
G-6-PD	Glucose-6-phosphate deshydrogénase	1,20	1,70		
6-PGD	6-phosphogluconate deshydrogénase	0,50	0,53		

Les activités érythrocytaires sont exprimées en μmoles de substrat transformé par min et par ml de globules rouges.
Celles des leucocytes, en μmoles par min et par g de protéines (dosées selon LOWRY).

Tableau II. Intermédiaires de la voie glycolytique (en μmoles par ml de globules rouges)

	Normaux	Témoin avec 20% de réticulocytes	Nadia Am..
Glucose-6-P	31,3	86	60,4
Fructose-6-P	9,7	24	15,1
Fructose-1,6-DP	3,8	–	6,0
Trioses P	19,2	14	12,7
3-P-glycérate	66,3	62	78,5
2-P-glycérate	11,8	38	15,1
P-énolpyruvate	17,4	38	33,2
Glucose-1,6-DP	97,4	105	134,5
2,3-DP glycérate	4582	4500	5547
ATP	1527	1736	1541
ADP	170	285	223
AMP	20	66	72
Σ (adényliques)	1717	2087	1836

cytes. Il convient de souligner que le déficit en PHI n'accumule pas de glucose-6-phosphate.

Le dosage des pyridines nucléotides montre (tableau III) : une diminution de ($NAD^+ + NADH$) et une augmentation de ($NADP^+ + NADPH$) ; une élévation du rapport $NADH/NAD^+$, alors que l'état redox des NADP est normal.

4. En gel d'amidon (tampon Tris/citrate pH 8,1), la PHI du malade se comporte comme celle des normaux [voir également les deux cas récemment rapportés par PAGLIA *et al.*, 7].

5. Les parents (d'origine tunisienne et cousins germains) ont un déficit en PHI supérieur à 50% chez le père, plus discret chez la mère (réticulocytose 4%). Les trois frères sont également déficients (50% environ). Chez le père et les trois frères, ce déficit ne s'accompagne d'aucune manifestation hématologique (tableau IV).

Tableau III. Teneur en pyridine-nucléotides (en nmoles par ml de globules rouges)

	Normaux	Nadia Am...
NAD^+	84,2 + 2,6	41,5
NADH	54,9 ± 1,8	64,6
(NAD^+ + NADH)	139,1 ± 5,1	106,1
$NADH/NAD^+$	0,65 ± 0,02	1,55
$NADP^+$	30,8 ± 1,5	46
NADPH	16,5 ± 0,9	26
($NADP^+$ + NADPH)	46,6 ± 2,0	72
$NADPH/NADP^+$	0,54 ± 0,03	0,56

Tableau IV. Enzymes glycolytiques de la famille de Nadia Am...

	Témoins		Père	Mère	Khalhed (frère)	Sami (frère)	Farid (frère)
Réticulocytose	0	6%	1%	4%	0,2%	0,2%	0,4%
Phosphohexo-isomérase	6,13	9,0	2,72	4,09	3,18	3,41	3,18
Glucose-6-P-deshydrogénase	1,20		1,13	1,82	1,36	1,13	1,19
Hexokinase	0,14		0,14	0,37	0,20	0,16	0,20
Pyruvate-kinase	2,70		2,27	3,75			
Phosphofructokinase	1,30		1,42	1,70			

Discussion

Cette observation de déficit important en PHI est très comparable à celle de BAUGHAN *et al.* [1]. Elle est intéressante sur le plan génétique, en raison de la consanguinité des parents, tous deux déficitaires. Elle pose cependant deux problèmes importants:

a) d'une part, le père et trois frères de notre malade, bien que déficients en PHI, ne présentaient aucune manifestation hématologique;

b) d'autre part, la relation entre la fragilité globulaire et les modifications métaboliques associées au déficit enzymatique reste inexplicable. Dans le déficit en pyruvate-kinase (enzyme régulateur de la glycolyse), la chute de l'ATP – par défaut de recharge – peut être incriminée dans le phénomène d'hémolyse. Dans notre cas, le déficit en PHI (enzyme non limitant) n'entraîne aucune modification au niveau des intermédiaires de la glycolyse et le taux d'ATP est normal. Seule la teneur en NAD érythrocytaire est diminuée et l'équilibre $NADH/NAD^+$ augmenté. OSKI et DIAMOND [8] ont signalé des modifications analogues dans une anémie par déficit en pyruvate kinase. Nous poursuivons actuellement cette exploration des systèmes redox de la glycolyse dans les anémies hémolytiques pour tenter de comprendre leur incidence sur la fragilité du globule rouge.

Summary

A case of hereditary hemolytic anemia with glucose phosphate isomerase deficiency is reported, in a 5-year-old girl. The defect is transmitted by an autosomal recessive mode of inheritance. The level of glycolytic intermediates is normal but there are modifications in the concentrations of pyridine nucleotides. The defect is also present in lymphocytes and granulocytes. In starch gel, electrophoretic migration of deficient enzyme is normal.

References

1. BAUGHAN, P.; VALENTINE, W.; PAGLIA, D.; WAYS, P.; SIMONS, E. and DEMARSCH, Q.: Hereditary hemolytic anemia associated with glucose phosphate isomerase (GPI) deficiency – a new enzyme defect of human erythrocytes. Blood *32:* 236–249 (1968).
2. CARTIER, P.: Dosage des pyridine nucléotides oxydés et réduits dans le sang et les tissus animaux. Europ. J. Biochem. *4:* 247–255 (1968).
3. CARTIER, P.; LEROUX, J. P. et MARCHAND, J. C.: Techniques de dosage des enzymes glycolytiques tissulaires. Ann. Biol. Clin. *25:* 109–136 (1967).
4. CARTIER, P.; LEROUX, J. P. et TEMKINE, H.: Techniques de dosage des intermédiaires de la glycolyse dans les tissus. Ann. Biol. Clin. *25:* 791–813 (1967).

5. CARTIER, P. et TEMKINE, H.: Le 2,3-diphosphoglycérate et le glucose-1,6-diphosphate du globule rouge: techniques de dosage. Ann. Biol. Clin. *25:* 1119–1128 (1967).
6. MARCHAND, J. C. et GARREAU, H.: L'activité des enzymes glycolytiques du réticulocyte et du globule rouge en fonction de l'âge. C. R. Soc. Biol. *162:* 1302–1306 (1968).
7. OSKI, F. A. and DIAMOND, L. K.: Erythrocyte pyruvate kinase deficiency resulting in congenital nonspherocytic hemolytic anemia. New England J. Med. *269:* 763–770 (1963).
8. PAGLIA, D. E.; HOLLAND, P.; BAUGHAN, M., and VALENTINE, W.: Occurrence of defective hexosphosphate isomerization in human erythrocytes and leucocytes. New England J. Med. *270:* 66–71 (1969).

Adresses des auteurs: Prof. P. CARTIER, Dr. H. TEMKINE et Dr. C. GRISCELLI, Enzymologie Médicale, INSERM, Unité 75, et Clinique Médicale Infantile, 156, rue de Vaugirard, *75 – Paris 15e* (France)

7th int. Congr. clin. Chem., Geneva/Evian 1969; vol. 2: Clinical Enzymology, pp. 144–149
(Karger, Basel/München/Paris/New York 1970)

The Possibility of the Existence of a Variety of Glucose-6-phosphate Dehydrogenase Peculiar to the Algerian Race: Glucose-6-phosphate Dehydrogenase Type Debrousse

C. KISSIN and J. COTTE
Laboratoire d'Enzymologie, Hôpital d'Enfants Debrousse, Lyon

The deficiency in glucose-6-phosphate dehydrogenase (G-6-PD) is probably the most prevalent genetic mutation known for the present time. Approximatively one hundred million people are affected by it in the world.

G-6-PD deficiency can be classified in two main groups, forming the majority of the cases and defined by clinical and biochemical criteria:

1. The first group consists of deficiences among Negroes with type A$^-$ enzyme.

2. The second group consists of deficiences in the white race which is type B$^-$, more commonly called Mediterranean type, since it is found all along the Mediterranean region (fig. 1).

Nevertheless, it is curious that races having apparently no common ancestors should have a pathological enzyme with similar properties. RAMOT has already shown among the Jews the existence of an electrophoretical mobility slightly different from the Mediterranean enzyme [7] and KIRKMANN, among the Greeks, showed the existence of an enzyme with an affinity for structural analogues, which was also different from it [4]. We have been concerned with the hypothesis of the existence of a variety of glucose-6-phosphate dehydrogenase peculiar to the Algerian race.

Some studies have shown that G-6-PD deficiency is also frequent in North Africa and particularly in Algeria [2], but no biochemical and electrophoretical studies have, as yet, been carried out among Algerians except by KAPLAN in a white Kabyle [3]. To confirm our hypothesis we studied G-6-PD of two Algerian deficient males belonging to unrelated families. We report here the discovery of a fast electrophoretic variant, G-6-PD Debrousse, with kinetic characteristics very different from those of the Mediterranean type.

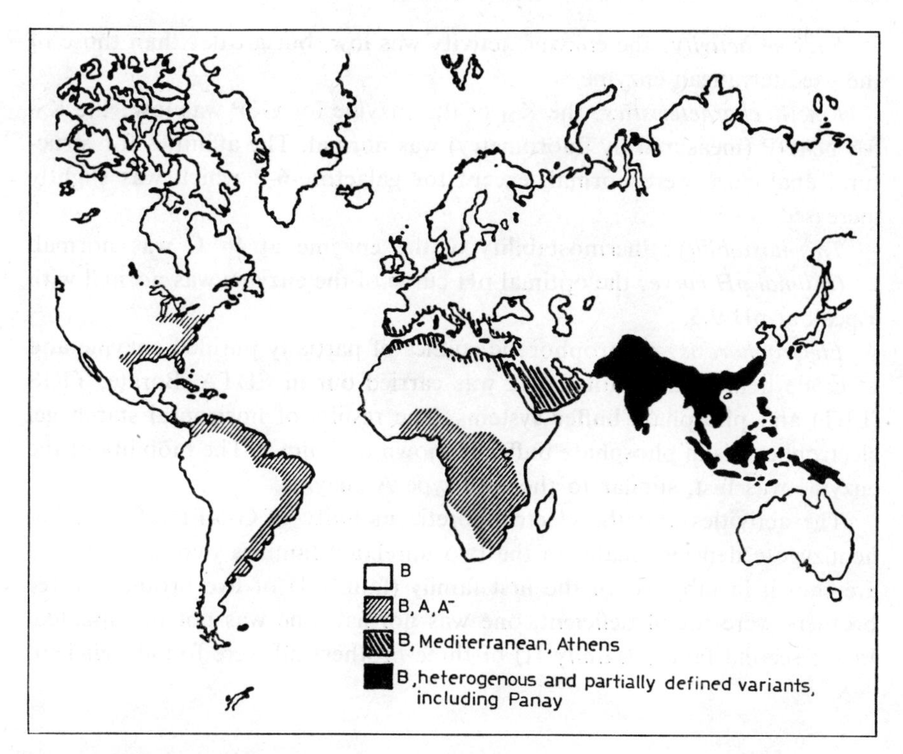

Fig. 1. World distribution of variants of G-6-PD according to M. N. FERNANDEZ and V. F. FAIRBANKS (1968).

Materials and Methods

The technique used was the standard one according to the World Health Organization [8] for the purification and characterization of G-6-PD from the erythrocytes of the two deficient males.

Mourad Ag..., the first one (family I), was a five-year-old child of native Algerian race, a descendent of several generations from the Constantine district [5].

Said T..., the second subject (family II), aged six, was also of native Algerian race, out of several generations from the El Oued district.

They never presented any clinical signs of hemolysis and their deficiencies were found during systematic studies.

Results

The results obtained are summarized in table I and consist mainly of the following:

Enzyme activity: the enzyme activity was low, but greater than those of the Mediterranean enzyme.

Kinetic characteristics: the K_M of the enzyme for G6P was low; the K_M for NADP (measured by fluorometry) was normal. The affinities for structural analogues were normal, except for galactose-6-P which was slightly increased.

Thermostability: thermostability of the enzyme at 46° C was normal.

Optimal pH curve: the optimal pH curve of the enzyme was normal with a peak at pH 9.5.

Electrophoresis: electrophoretic studies of partially purified enzyme and of crude hemolysate diluted 1:2 was carried out in EDTA, Borate, TRIS (EBT) and phosphate buffer systems. The results of horizontal starch gel electrophoresis in phosphate buffer is shown in figure 2. The mobility of the enzyme was fast, similar to those of type A enzyme.

The activities and the electrophoretic mobility of G-6-PD of the other hemizygote deficient males in the two unrelated families were studied and are shown in table II. In the first family (family I) of five brothers, three brothers were found deficient, one was normal, one was not investigated. In the second family (family II) of three brothers all were found deficient.

Table I. Comparison of G-6-PD B^+ (normal), G-6-PD B^- (Mediterranean), and G-6-PD Debrousse

	G-6-PD Debrousse	G-6-PD B^+ (normal)	G-6-PD B^- (Mediterranean)
RBC activity	20	100	0–7
Electrophoretic mobility	110	100	100
Michaelis constants ($K_M \times 10^{-6}$) for			
G6P	19–29	52–78	19–26
NADP	2,5 (fluorimetry)	2,9–4,4	1,2–1,6
Substrate utilization (% of G6P rate)			
Galactose-6-P	9	4	20–35
2-deoxy-G6P	4	4	23–37
Glucose (5)	4	3	10–30
Optimal pH	truncate	truncate	biphasic
Thermostability	normal	normal	decreased

Table II. Studies in the deficient males of the two affected families

Family	Origin	Subject	Relation	G-6-PD activity in RBC (% normal)	Electrophoretic mobility
I	Constantine	Nasser Ag.*	sib.	5	110
		Mohamed Ag.	sib.	16	110
		Fari Ag.	sib.	37	110
II	El Oued	Youssef T.	sib.	9	110
		Nasser T.	sib.	37	110

* G-6-PD of Nasser Ag. was purified and studied. It showed the same biochemical characteristics as G-6-PD Debrousse.

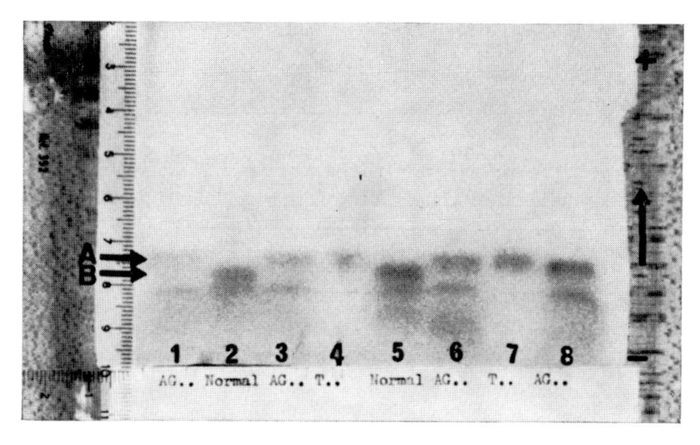

Fig. 2. Starch gel electrophoresis of G-6-PD Debrousse. Electrophoresis was carried out for 14 h at 4°C at 4V /cc with phosphate buffer (pH 7). G-6-PD Debrousse is in slots 1, 3,4, 6, 7, 8; G-6-PD B^+ is in slots 2 and 5.

The enzyme activities were found to lie between 5 and 37% with an average of 18% in one family and 24% in the other. The electrophoretic mobility was the same for all, i.e. fast, – similar to those of type A –; the mothers showed two bands, one with normal mobility and one with fast mobility. The enzyme of a brother of one case (Nasser Ag...) was isolated, purified and studied and it showed the same biochemical characteristics as G-6-PD Debrousse.

Discussion

The similarities and differences of G-6-PD Debrousse with normal enzyme and Mediterranean enzyme are shown in table III.

It can be seen that there is only one similarity with the Mediterranean enzyme, i.e. the low K_M for G6P. G-6-PD Debrousse differs also from all other known variants [1]. These studies appear to confirm the hypothesis of an enzyme very different from the Mediterranean enzyme and peculiar to the Algerian race.

Table III

Comparison of G-6-PD Debrousse and Normal (B^+)	
Similarities	Differences
K_M for NADP	RBC activity
2-deoxy-G6P utilization	Electrophoretic mobility
Glucose utilization	K_M for G6P
Optimal pH	Galactose-6-P utilization
Thermostability	

Comparison of G-6-PD Debrousse and Mediterranean enzyme (B^-)	
Similarities	Differences
K_M for G6P	RBC activity
	Electrophoretic mobility
	K_M for NADP
	Galactose-6-P utilization
	2-deoxy-G6P utilization
	Glucose utilization
	Optimal pH
	Thermostability

Summary

A new variant of G-6-PD found in 2 Algerian kindreds belonging to unrelated families is reported. This variant (G-6-PD Debrousse) has an activity of about 20% of the normal. The kinetic characteristics investigated are normal except the K_M for G6P and the electrophoretic mobility is fast-like type A. It is, therefore, very different from the Mediterranean type and appears to confirm the possibility of an enzyme peculiar to the Algerian race.

References

1. Beutler, E.: Drug induced hemolytic anemia. Pharmacol. Rev. *21:* 73 (1969).
2. Messerschmitt, J.; Suaudeau, C.; Benallegue, A.; Venezia, R.; Fabre, S.; Bon, J.; Andre, L.; Khati, B.; Dubois, M.; Benabdallah, S. et Kotchoyan, P.: Défaut en G6PD et anémies hémolytiques en Algérie. Nouv. Rev. franç. Hémat. *7:* 827 (1967).
3. Kaplan, J. C.; Rosa, R.; Seringue, P. et Hoeffel, J. C.: Le polymorphisme de la G-6-PDH chez l'homme. II. Etude d'une nouvelle variété à activité diminuée: le type Kabyle. Enzymol. biol. clin. *8:* 332 (1967).
4. Kirkmann, H. N.; Doxiadis, S. A.; Valaes, T.; Tasso-Poulos, N. and Brinson, A. G.: Diverse characteristics of G6PD from Greek children. J. Lab. clin. Med. *65:* 212 (1965).
5. Kissin, C. and Beutler, E.: The utilization of glucose by normal glucose-6-phosphate dehydrogenase and by glucose-6-phosphate dehydrogenase Mediterranean. Proc. Soc. exp. Biol. Med. *128:* 595 (1968).
6. Kissin, C. et Cotte, J.: Etude d'un variant de glucose-6-phosphate déshydrogénase: le type Constantine. Enzymol. biol. clin. *11:* 277 (1970).
7. Ramot, B.; Bauminger, S.; Brok, F.; Gafni, D. and Schwartz, J.: Characterisation of G6PD in Jewish mutants. J. Lab. clin. Med. *64:* 895 (1965).
8. World Health Organisation: Standardisation of procedures for the study of G6PD. WHO. Techn. Rep. Ser. *338:* 1 (1969).

Authors' address: Dr. Christiane Kissin and Dr. J. M. Cotte, Laboratoire d'Enzymologie, Hôpital d'Enfants Debrousse, *F-69 Lyon 5e* (France)

7th int. Congr. clin. Chem., Geneva/Evian 1969; vol. 2: Clinical Enzymology, pp. 150–154
(Karger, Basel/München/Paris/New York 1970)

Proteinchemische und enzymatische Analyse von Pankreascystensaft[1]

F. Willig, F. H. Schmidt und H. Stork

I. Medizinische Universitätsklinik Mannheim und Forschungsabteilung Boehringer, Mannheim

Pseudocysten der Bauchspeicheldrüse gelten in gewissem Prozentsatz als Komplikation von verschiedenen Pankreatitisformen. Sie entstehen aus Relikten von Gewebsnekrosen, die abgebaut worden sind und einen intraparenchymatösen Hohlraum hinterlassen haben [2]. Sie sind häufig von einem blutigen, manchmal auch infizierten Material erfüllt. Seltener sind sie gereinigt, kommunizieren intermittierend mit dem Bauchspeicheldrüsengangsystem und enthalten mehr oder weniger reines Pankreasexkret.

Wir hatten Gelegenheit, bei einer 33jährigen Frau im Anschluß an eine akute, nekrotisierende Pankreatitis die Entwicklung einer Pseudocyste zu beobachten. Bei der Operation konnten wir ca. 600 ml sterile, klare, bernsteinfarbene Flüssigkeit gewinnen. Diese war mit einem pH von 8,2 leicht alkalisch. Da sich die Entwicklung der Pseudocyste über mehrere Monate ausgedehnt hat, interessierte uns, ob in der Flüssigkeit überhaupt aktive Enzyme enthalten waren. Weiterhin interessierte der Vergleich mit den bekannten Daten des nach Reiz gewonnenen Pankreasexkrets.

Enzymatische Untersuchungen

Die Amylaseuntersuchung zeigte, daß es sich um einen enzymatisch aktiven Saft handelte. Die Höhe der Aktivität konnte durchaus der eines normalen Pankreassaftes entsprechen. Ähnliche quantitative Verhältnisse zeigten sich auch bei der Bestimmung der Lipaseaktivität (Tab. I). Besonders wichtig erschien uns die Bestimmung der proteolytischen Enzyme, da

[1] Herrn Prof. Dr. W. Hoffmeister zum 60. Geburtstag gewidmet.

Tabelle I. Pankreascystenflüssigkeit

	Substrat		
Proteasen	Azocasein		
(spontan)		<0,1	μg/ml
(nach Enterokinase)		112	μg/ml
Trypsin	BAPA	42	μg/ml
Chymotrypsin	SUPHEPA	58	μg/ml
AS-Arylamidase (LAP)	LEUPA	1,8	mU/ml
Amylase	Stärke		
(amyloklastisch)		71 000	mU/ml
(saccharometrisch)		12 000	mU/ml
Lipase (titrimetrisch)	Olivenöl	175	U/ml
Natrium		118	mval/l
Kalium		3,0	mval/l
Calcium		0,9	mval/l
Chlorid		79	mval/l
Gesamtprotein		2,42	mg/ml

diesen – in ihrer aktiven Form – immer wieder eine wesentliche pathogenetische Bedeutung für den Ablauf einer Pankreatitis zugeschrieben wurde. Die Gesamtproteolyse bestimmten wir mit Hilfe des Substrats Azocasein. Trypsin wurde mit N-α-Benzoyl-DL-arginin-p-nitroanilid (BAPA), Chymotrypsin mit N-Succinyl-L-phenylalanin-p-nitroanilid (SUPHEPA) nachgewiesen [6]. Weiterhin dienten uns die Substrate L-Leucin-p-nitroanilid, α-Glutaminyl-p-nitroanilid und γ-Glutaminyl-p-nitroanilid zur Erfassung von Aminosäure-arylamidasen. Wir nahmen die Untersuchungen dieser Aktivitäten vor, weil insbesondere der LAP früher eine gewisse Pankreasspezifität zugeschrieben wurde. Weder mit dem Leucin- noch mit dem Glutaminylsubstrat fanden sich nennenswerte Aktivitäten. Damit kann, wie bereits früher gezeigt, die Bestimmung dieser Enzyme im Serum nicht als pankreasspezifisch gelten [9, 10].

Die spontane proteolytische Aktivität ist minimal. Nach Aktivierung *in vitro* mit Enterokinase findet sich eine hohe, normalem Pankreassaft entsprechende Proteolyse. Die Addition von tryptischen und chymotryptischen Aktivitäten ergibt nur einen Teil der Gesamtproteolyse. Dieser Befund weist auf die Aktivierung weiterer, nicht mit Trypsin und Chymo-

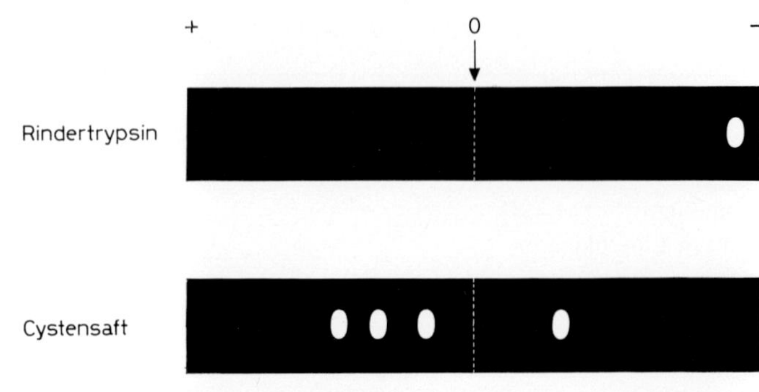

Abb. 1. Agarose-Gel-Elektophorese, Proteasen, Substrat = Casein

trypsin identischer Proteasen hin. Es zeigt sich tatsächlich (Abb. 1), daß bei einer elektrophoretischen Auftrennung und anschließender topochemischer Charakterisierung mit Casein als Substrat mindestens vier Proteine nach ihrer Wanderungsgeschwindigkeit im elektrischen Feld unterschieden werden können, die proteolytische Aktivität aufweisen. Die anodischen Proteasen entsprechen [4, 5, 7] offenbar verschiedenen Enzymen mit tryptischer bzw. chymotryptischer Spezifität. Die kathodische Spaltung kommt dagegen mit hoher Wahrscheinlichkeit einem elastolytischen Enzym zu.

Das Elektrolytspektrum entspricht in seinen Relationen dem des Bauchspeichels (Tab. I) und unterscheidet sich eindeutig vom Mundspeichel, dessen Kalium-Natriumquotient [3] wesentlich höher ist.

Proteinchemische Untersuchungen

Die Gesamtproteinbestimmung ergab mit der Biuretreaktion 2,42 mg/ml. Dies ist ein Wert wie er ebenfalls im physiologischen Exkret gefunden werden kann. Bei der elektrophoretischen Fraktionierung der konzentrierten Flüssigkeit ergaben sich folgende Relationen (Abb. 2): *A* – mit einer Wanderungsgeschwindigkeit von Serumalbumin – 46%; *B* 27%, *C:* 27%. Die weitere Analyse erfolgte mit immunchemischen Methoden. Zur Immunelektrophorese dienten uns handelsübliche Antihumanseren sowie Antiseren, die wir durch Sensibilisieren von Kaninchen mit unserem Pankreascystensaft gewonnen haben. Dabei entsprachen (Abb. 3) A Albumin, α_1-Antitrypsin und saures α_1-Glykoprotein. B korrelierte mit Transferrin und C mit Immunglobulin G. Dieser Befund unterscheidet den Cystensaft und den Bauchspeichel wesentlich vom Mundspeichel, in dem bekanntlich nur eine geringe Konzentration von Immunglobulin G, dagegen eine höhere von Immunglobulin A gefunden wird [1]. Diese Proteine konnten auch mit der radiären Immunodiffusionsmethode identifiziert werden. Bemerkenswert erscheint die Exkretion von α_1-Antitrypsin, das von uns auch in anderen Pankreasfistelsäften nachgewiesen werden konnte. Daß das Pankreasexkret Inhibitoren enthält, ist bekannt. α_1-Antitrypsin ist sicher nicht identisch mit dem von WERLE *et al.* [8] nachgewiesenen niedermolekularen Inhibitor aus menschlichem Pankreasexkret. Es könnte aber durchaus dem höhermolekularen Proteaseninhibitor entsprechen. Seine Konzentration (weniger als 5 mg/ml) reicht allerdings zu einer völligen Hemmung der Zymogenaktivierung nicht aus.

Mit Anticystensaftserum (APS) vom Kaninchen zeigten sich in der Immunelektrophorese weitere Präzipitate (Abb. 3), die nach der Wanderungsgeschwindigkeit und bei dem quantitativ hohen Anteil enzymatisch aktiver Proteine durchaus Pankreasenzymen entsprechen könnten. Es kommen u.a. in Frage: Trypsinogen, Chymotrypsinogen, Lipase, Elastase und Amylase.

Saft aus einer Pseudocyste des Pankreas erweist sich in seinen Bestandteilen als sehr ähnlich dem normalen Bauchspeichel, dessen Protein-, Elektrolyt- und Enzymzusammensetzung er aufweist. Er unterscheidet sich von jenem durch die fehlende Aktivierung der Proteasen. Vom Mundspeichel unterscheidet ihn das typische Elektrolytmuster und die Ausscheidung von Immunglobulin G anstelle von Immunglobulin A.

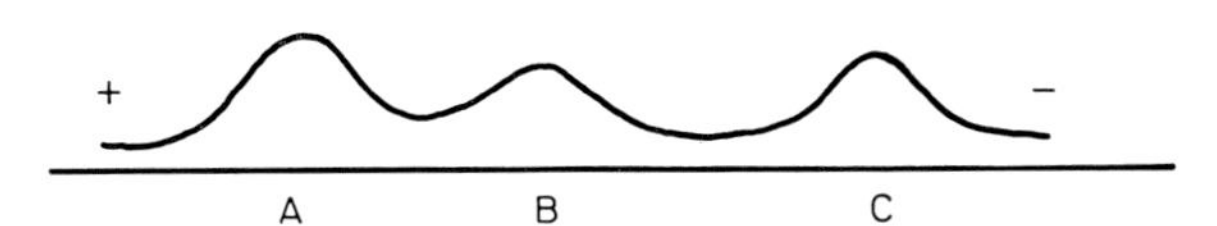

Abb. 2. Pankreascystensaft, Protein. 2,42mg/ml, Papierelektrophorese

Summary

Juice of a human pancreatic pseudocyst has been examined by several methods (see table I) including immunoelectrophoresis. Among other data the following results were obtained: There was no spontaneous proteolytic activity. After activation by enterokinase at least 4 proteases could be identified (e.g. trypsin, chymotrypsin). Besides few 'pancreatogenic' proteins, the following serum-proteins are established by immunoelectrophoresis: albumin, α_1-antitrypsin, transferrin, immunoglobulin G. There are typical differences between pancreatic juice and saliva: e.g. potassium-sodium-ratio and excretion of IgG instead of IgA.

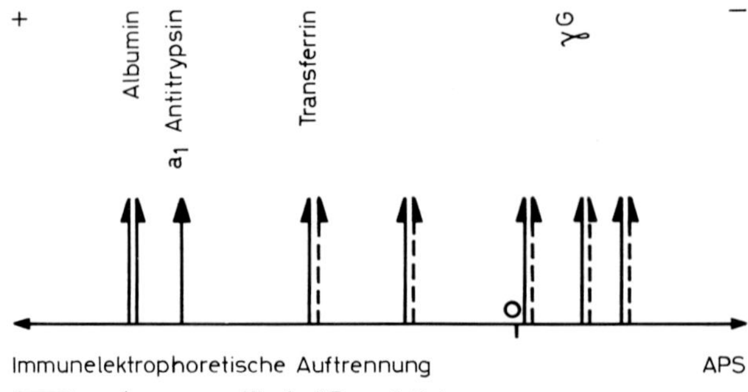

Abb. 3.

Literatur

1. GITLIN, D.: Current aspects of the structure, function, and genetics of the immunglobulins. Ann. Rev. Med. *17:* 1 (1966).
2. HESS, W.: Die chronische Pankreatitis. Klinik, Diagnostik und chirurgische Therapie der chronischen Pankreopathien (Huber, Bern/Stuttgart 1969).
3. HOFFMEISTER, W. und ALBRECHT, H.: Die Speichelelektrolyte und ihre diagnostische Bedeutung. Untersuchungen am Menschen unter besonderer Berücksichtigung endokriner Störungen. Klin. Wschr. *31:* 567 (1953).
4. KELLER, P. J. and ALLAN, B. J.: The protein composition of human pancreatic juice. J. biol. Chem. *242:* 281 (1967).
5. MARCHIS-MOUREN, G.: Etude comparée de l'équipement enzymatique du suc pancréatique de diverses espèces. Bull. Soc. Chim. biol. *47:* 2207 (1965).
6. NAGEL, W.; WILLIG, F.; PESCHKE, W. und SCHMIDT, F. H.: Über die Bestimmung von Trypsin und Chymotrypsin mit Aminosäure-p-nitroaniliden. Hoppe-Seyler's Z. Physiol. Chem. *340:* 1 (1965).
7. SILBERBERG, V. L. and HADORN, B.: Identification of pancreatic enzymes in human duodenal contents. Biochem. biophys. Acta *167:* 616 (1968).
8. VOGEL, R.; TRAUTSCHOLD, I. und WERLE, E.: Natürliche Proteinaseninhibitoren (Thieme, Stuttgart 1966).
9. WILLIG, F.: Pankreas- und Gallenwegserkrankungen. Pathophysiologische und klinische Untersuchungen zur Bedeutung der Proteolyse; Habil.-schrift Heidelberg-Mannheim (1968).
10. WILLIG, F.; GREINER, I.; STORK, H. und SCHMIDT, F. H.: Leucinaminopeptidase-(Arylamidase-) Aktivität im Serum. Bestimmung mit Leucin-p-nitroanilid als Substrat. Klin. Wschr. *45:* 474 (1967).

Adresse des Autors: Priv.-Doz. Dr. F. WILLIG, I. Medizinische Universitätsklinik, Postfach 23, *68 Mannheim* (Deutschland)

7th int. Congr. clin. Chem., Geneva/Evian 1969; vol. 2: Clinical Enzymology, pp. 155–162 (Karger, Basel/München/Paris/New York 1970).

Relations entre le taux des protéines totales et l'activité des enzymes dans le suc duodénal humain normal et pathologique

A. Ribet, J. P. Pascal, N. Vaysse[1], D. Augier et J. P. Thouvenot
Groupe de Recherches de Pathologie digestive, Clinique médicale Sud (Prof. J. Gadrat), Hôpital Purpan et Laboratoire de Biochimie (Prof. Douste-Blazy), Faculté de Médecine de Rangueil, Toulouse

Plusieurs études déjà anciennes ont examiné le problème des rapports des enzymes entre eux d'une part, et entre les enzymes et le taux de protéines d'autre part dans le suc pancréatique. L'école de Komarov en particulier [6, 7, 9] par plusieurs travaux chez le chien, a montré que la répartition des protéines enzymatiques est relativement constante chez le même individu, et qu'elle varie faiblement dans les différents échantillons de suc au cours d'un même test, même si la stimulation est répétée par des moyens differénts.

On sait pourtant que les enzymes ne sont pas inexorablement liés entre eux tant au cours du processus de synthèse et de migration intra-cellulaire qu'au cours de leur expulsion dans les canaux pancréatiques. En effet, sous l'influence de variations de la composition du régime, on peut obtenir des modifications de la composition enzymatique du tissu et du suc pancréatique [3, 5, 11], qui correspondent vraisemblablement à des variations de la vitesse de synthèse de certains enzymes [13].

Ces phénomènes n'ont cependant pas pu être reproduits chez l'homme [17]. Par contre, on peut provoquer l'apparition de telles modifications dans le suc duodénal humain en variant le type de stimulation [18], ou en prolongeant suffisamment une perfusion continue d'une forte dose de sécrétine [16].

Au cours d'un travail antérieur [15] nous avions fait des constatations semblables. Dans l'étude présente, nous avons repris ce problème, en l'élargissant à un plus grand nombre d'enzymes et en comparant les résultats chez les sujets témoins à ceux obtenus chez des sujets atteints de pancréatite chronique.

[1] Attachée de Recherche à l'INSERM

Matériel et méthodes

Sujets

Cette étude porte sur 35 sujets adultes des deux sexes, répartis de façon suivante:

1. Le groupe 'témoin' réunit les diagnostics suivants: dystonie neurovégétative à expression digestive (13 cas), éthylisme chronique avec gastrite, parotidite ou cirrhose (5 cas), cancer non digestif (2 cas), ulcère gastrique (1 cas), lithiase vésiculaire (1 cas), stéatorrhée idiopathique (1 cas), hyperlipémie (1 cas), glomérulopathie chronique (1 cas).

2. Le deuxième groupe comprend 10 pancréatites chroniques alcooliques, dont le diagnostic fut fait 5 fois par la présence de calcifications, trois fois en l'absence de calcifications par l'intervention chirurgicale et deux fois chez des pancréatites chroniques non calcifiantes par un ensemble clinique, radiologique et biologique typique.

Techniques

Après mise en place des sondes [14] et lorsque le liquide duodénal devient alcalin, une perfusion intraveineuse de sérum physiologique est mise en place dans une veine du pli du coude (200 ml/h). Elle sert de véhicule à la sécrétine [2]: 3 unités/kg de poids/h pendant les deux heures du test. De plus, de la 60e à la 120e minute est ajoutée à la sécrétine une perfusion de pancréozymine [2]: 1 unité ou 2 unités/kg de poids/h, selon les sujets.

Le suc est recueilli par échantillons de 10 min maintenus à 0° C. Dans les échantillons, on dose:

a) les protéines totales (16 sujets), selon la méthode de LOWRY [10] adaptée à l'autoanalyseur Technicon (2);

b) la lipase (35 sujets) sur une émulsion d'huile d'olive [19];

c) la trypsine (31 sujets) sur BAEE et la chymotrypsine sur ATEE avec dosage titrimétrique [4];

d) l'amylase (18 sujets) par une méthode amyloclastique avec dosage de l'amidon non hydrolysé par le test colorimétrique à l'iode [1].

e) la phospholipase A (8 sujets) par isolement chromatographique sur couche mince des produits de la réaction enzymatique et évaluation spectrophotométrique du phosphore des phospholipides dégradés. Les valeurs sont exprimées par le rapport lysolécithines formées/lécithines.

Résultats

1. Relations entre enzymes chez les sujets témoins

La figure 1 illustre l'aspect des relations entre enzymes. La chymotrypsine n'est pas représentée car son comportement est tout à fait semblable à

[2] Boots Pure Drug Cy. Ltd., Nottingham, Angleterre.

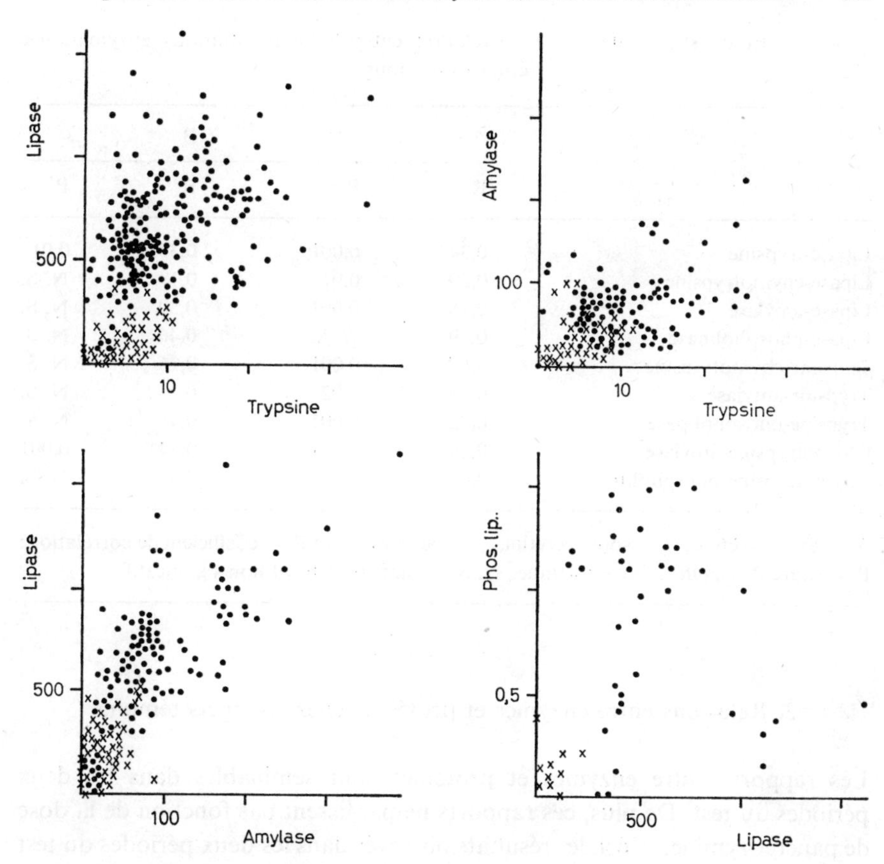

Fig. 1. Relations entre certaines activités enzymatiques.
O: sujets témoins
×: sujets atteints de pancréatite chronique.

celui de la trypsine. Le tableau I montre que, sous stimulation par sécrétine seule, les variations entre enzymes se font de façon parallèle. Par contre, il n'y a pas de corrélation statistiquement significative lorsque le suc est recueilli sous perfusion de sécrétine + pancréozymine. Ceci est vraisemblablement dû au fait que dans l'étude statistique ont été groupés des sujets stimulés par 1 et par 2 unités/kg/h de pancréozymine. En effet, lorsque l'on sépare ces deux populations, on voit qu'elles se répartissent de façon différente et que les rapports paraissent meilleurs lorsque la dose de pancréozymine est forte (fig. 2).

Tableau I. Etude statistique de la corrélation entre les concentrations enzymatiques étudiées par jour

	A		B	
	R	P	R	P
Lipase-trypsine	0,347	0,001	0,335	0,01
Lipase-chymotrypsine	0,29	0,01	0,19	N. S.
Lipase-amylase	0,78	0,001	0,30	N. S.
Lipase-phospholipase	0,19	N. S.	0,46	N. S.
Trypsine-chymotrypsine	0,67	0,001	0,26	N. S.
Trypsine-amylase	0,43	0,02	0,24	N. S.
Trypsine-phospholipase	0,72	0,001	0,48	N. S.
Chymotrypsine-amylase	0,89	0,001	0,84	0,001
Chymotrypsine-phospholipase	0,08	N. S.	0,43	N. S.

A = sous sécrétine; B = sous sécrétine + pancréozymine; R = coefficient de corrélation; P = degré de signification statistique; N. S. = statistiquement non significatif.

2. Relations entre enzymes et protéines chez les sujets témoins

Les rapports entre enzymes et protéines sont semblables dans les deux périodes du test. De plus, ces rapports ne paraissent pas fonction de la dose de pancréozymine. Ainsi, les résultats observés dans les deux périodes du test chez tous les sujets témoins ont-ils été finalement groupés pour cette étude.

Il existe une excellente corrélation entre le taux de chaque enzyme et le taux de protéines (tableau II). Cette corrélation est particulièrement claire pour l'amylase et pour la lipase.

3. Relations entre enzymes et protéines chez les sujets atteints de pancréatite chronique

Il est visible dans la figure 3 qu'elles sont différentes de celles observées chez les témoins. On peut constater que, dans l'ensemble, les taux d'enzymes sont plus abaissés que les taux de protéines totales. Ce phénomène est inconstant pour la trypsine et la chymotrypsine, mais il est extrêmement net pour la lipase et l'amylase.

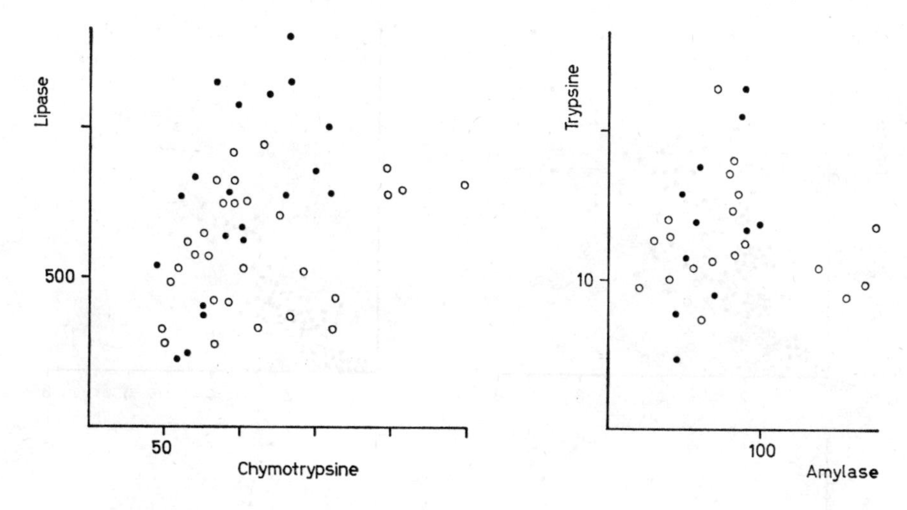

Fig. 2. Relations entre certaines activités enzymatiques sous perfusion de sécrétine + pancréozymine.

○: 1 U/Kg/H de pancréozymine.

●: 2 U/Kg/H de pancréozymine.

Tableau II. Etude statistique des relations entre les concentrations de protéines totales et les activités enzymatiques

Protéines et:	N	R	P
Lipase	128	0,727	0,001
Trypsine	84	0,476	0,001
Chymotrypsine	84	0,470	0,001
Phospholipase	32	0,543	0,05
Amylase	60	0,910	0,001

N = nombre d'échantillons; R = coefficient de corrélation; P = degré de signification statistique.

Discussion

Nos résultats ne se superposent pas à ceux exposés dans des travaux antérieurs [8, 15, 18, 21]. En effet, les rapports entre enzymes n'ont pû être analysés correctement sous stimulation de pancréozymine du fait des deux doses employées. On peut cependant noter que la dose de pancréozymine

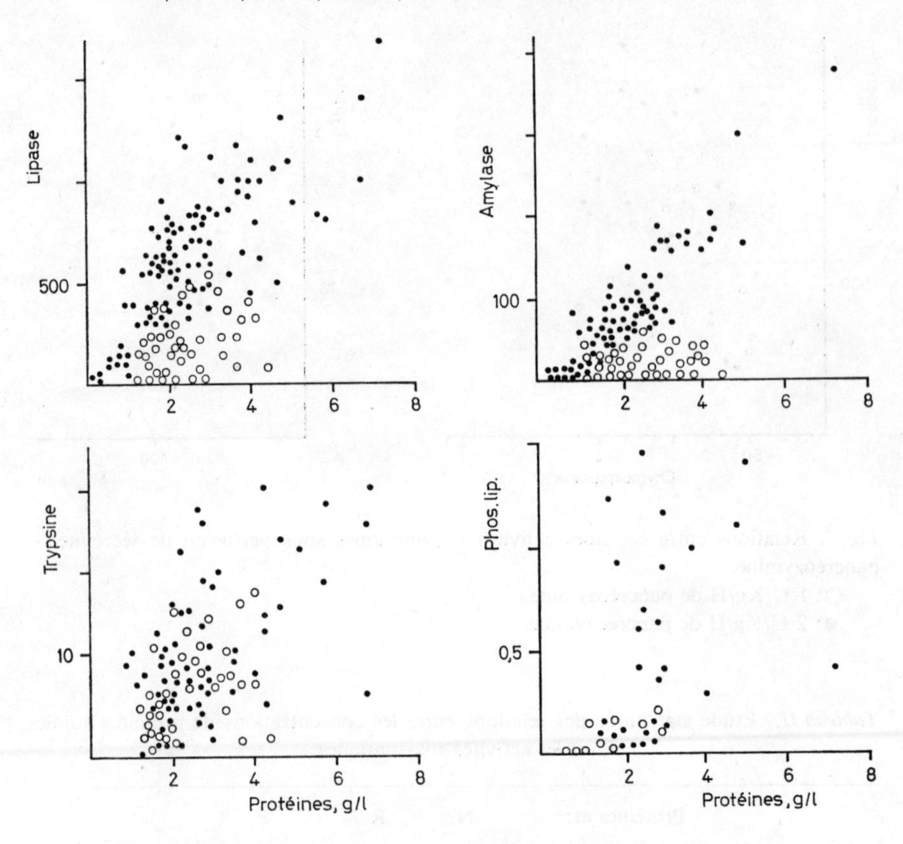

Fig. 3. Relations entre le taux de protéines totales et certaines activités enzymatiques.
●: sujets témoins.
○: sujets atteints de pancréatite chronique.

influence visiblement la répartition enzymatique dans le suc pancréatique. D'autre part, malgré la dégradation enzymatique partielle [12] qui intervient dans le suc préformé sécrété sous sécrétine, les rapports entre enzymes sont relativement constants.

Chez les sujets atteints de pancréatite chronique, il est remarquable de constater le faible abaissement du taux de protéines totales du suc et la modification de la répartition des enzymes qui comporte un faible abaissement des enzymes protéolytiques et un effondrement de la lipase et de l'amylase. Ces constatations peuvent être expliquées par le fait que ces deux derniers enzymes ne représentent qu'une faible proportion des protéines du suc pancréatique. On peut en conclure que, pour le diagnostic d'une insuffisance pancréatique, le dosage du taux de protéines est dénué d'intérêt. Les dosages de la trypsine et de la chymotrypsine sont infidèles du fait de la dégradation enzymatique spontanée qui abaisse la limite inférieure des valeurs normales.

A l'opposé, les dosages de l'amylase et de la lipase sont de loin les plus fidèles, dans les conditions réalisées ici. Ces deux enzymes varient d'ailleurs généralement de façon étroitement parallèle. Le nombre de mesures est insuffisant pour avoir une opinion sur la valeur diagnostique du dosage de l'activité phospholipasique. Chez les sujets témoins par contre, le dosage des protéines totales donne une image de la sécrétion enzymatique et suffit donc pour les études physiologiques.

Summary

The relationships between the total protein content and several enzyme activities (lipase, amylase, trypsin, chymotrypsin and phospho-lipase) were studied in the duodenal aspirate of 35 subjects (25 control subjects and 10 chronic pancreatitis). It was observed that the enzyme activities were well correlated with the protein content in control subjects. However, in patients with chronic pancreatitis the enzyme activities were grossly diminished, while the total protein content was not significantly modified. It is concluded that the measurement of total protein concentration gives a good idea of the enzyme content of duodenal juice in normal persons although the measurement of one or several enzyme activities (especially lipase and amylase) is necessary for the diagnosis of chronic pancreatitis.

Bibliographie

1. Bourse, R.; Vaysse, N. et Augier, D.: Dosage automatique de l'amylase dans le liquide duodénal. Rev. Méd., Toulouse (à paraître).
2. Chariot, J.; Roze, C. et Souchard, M.: Dosage simultané des protides totaux et de l'amylase sur des micro-échantillons de suc pancréatique de rat. Symp. Technicon Europ. Automation en chimie analytique, pp. 383–387, Paris 1966.
3. Desnuelle, P.; Reboud, J. P. and Ben Abdeljlil, A.: Influence of the composition of diet on the enzyme content of rat pancreas; in Ciba Found. Symp. Exocrine Pancreas, pp. 90–107 (Churchill, London 1962).
4. Figarella, C.; Taulier, J. et Sarles, H.: Dosage de la chymotrypsine et de la trypsine dans le suc duodénal. Bull. Soc. Chim. biol. *47:* 679–686 (1965).
5. Grossman, M. I.; Greengard, H. and Ivy, A. C.: The effect of dietary composition on pancreatic enzymes. Amer. J. Physiol. *138:* 676–682 (1943).

6. GUTH, P. H.; KOMAROV, S. A.; SHAY, H. and STYLE, C. Z.: Relationship between protein nitrogen, proteolytic, amylolytic and lipolytic enzymes in canine pancreatic juice obtained under various conditions of stimulation. Amer. J. Physiol. *187:* 207–223 (1956).
7. GUTH, P. H.; KOMAROV, S. A.; SHAY, H. and STYLE, C. Z.: Pancreatic lipase determination in canine pancreatic secretion and human blood serum: Relationship between lipase and protein nitrogen in canine pancreatic juice secreted in response to different test meals and under different dietary regimens. Amer. J. Physiol. *192:* 1–13 (1958).
8. KELLER, P. J.; COHEN, E. and NEURATH, H.: The proteins of bovine pancreatic juice. J. biol. Chem. *233:* 344–349 (1958).
9. KOMAROV, S.; SIPLET, H.; SHAY, H. A. and STYLE, C. Z.: Relation of proteolytic and lipolytic activity of canine pancreatic secretion to protein nitrogen content. Fed. Proc. *13:* 81 (1954).
10. LOWRY, O. H.; ROSEBROUGH, N. J.; FARR, A. L. and RANDALL, R. J.: Protein measurement with the Folin phenol reagent. J. biol. Chem. *193:* 265–275 (1951).
11. MAGEE, D. F. and ANDERSON, E. G.: Changes in pancreatic enzymes brought about by alteration in nature of dietary protein. Amer. J. Physiol. *181:* 78–82 (1955).
12. PETERSON, H.; MYREN, J. and FOSS, O. P.: Total nitrogen content of human duodenal juice before and after intravenous injection of secretin. Scand. J. Gastroent. *2:* 1–6 (1967).
13. REBOUD, J. P.; MARCHIS-MOUREN, G.; PASERO, L.; COZZONE, A. et DESNUELLE, P.: Adaptation de la vitesse de biosynthèse de l'amylase pancréatique et du chymotrypsinogène à des régimes riches en amidon ou en protéines. Biochim. biophys. Acta *117:* 351–367 (1966).
14. RIBET, A.; PASCAL, J. P. et SANNOU N.: Etude de la fonction exocrine du pancréas humain par les perfusions continues de sécrétine. I. Influence des doses croissantes sur la sécrétion hydroélectrolytique. Arch. Mal. Appar. digest. *56:* 677–684 (1967).
15. RIBET, A.; PASCAL, J. P. et VAYSSE, N.: Recherche de la capacité enzymatique maxima du pancréas humain normal sous perfusion continue de sécrétine et de pancréazymine. Biol. Gastroent. *2:* 163–170 (1968).
16. SARLES, H.; BAUER, J. B. et PREZLIN, G.: Etude des injections répétées et des perfusions continues de sécrétine chez l'homme. Arch. Mal. Appar. digest *54:* 177–194 (1965).
17. SARLES, H.; FIGARELLA, C. et SOUVILLE, C.; L'exploration du pancréas par le tubage duodénal. Vie méd. *47:* 563–568 (1966).
18. SARLES, H.; FIGARELLA, C.; PREZLIN, G. et SOUVILLE, C.: Comportement différent de la lipase, de l'amylase et des enzymes protéolytiques après différents modes d'excitation du pancréas humain. Bull. Soc. Chim. biol. *48:* 951–957 (1966).
19. SARLES, H.; TAULIER, J. et FIGARELLA, C.: Dosage de la lipase dans le suc duodénal Rev. franç. Et. clin. biol. *8:* 706–707 (1963).
20. THOUVENOT, J. P. et DOUSTE-BLAZY, L.; Détermination des activités phospholipasiques du suc duodénal. Ann. Biol. clin. (à paraître).
21. WORNING, H. and MULLERTZ, S.: pH and pancreatic enzymes in the human duodenum during digestion of a standard meal. Scand. J. Gastroent. *1:* 268–283 (1966).

Adresses des auteurs: Dr. A. RIBET, Dr. J. P. PASCAL, Dr. N. VAYSSE, Dr. D. AUGIER et Dr. J. P. THOUVENOT, Groupe de Recherches de Pathologie digestive, Clinique médicale Sud, Hôpital Purpan, et Laboratoire de Biochimie, Faculté de Médecine de Rangueil, *31 - Toulouse* (France)

7th int. Congr. clin. Chem., Geneva/Evian 1969; vol. 2: Clinical Enzymology, pp. 163–172
(Karger, Basel/München/Paris/New York 1970)

Etude critique des tests de détection et d'identification des erreurs innées du métabolisme du glycogène (à propos de 17 cas de glycogénoses)

M. MATHIEU[1] et J. COTTE
Laboratoire d'Enzymologie, Hôpital d'Enfants Debrousse, Lyon

Introduction

Les glycogénoses représentent un pourcentage relativement élevé des erreurs innées du métabolisme susceptibles d'être détectées dans la première enfance. Ce sont les plus fréquents des troubles congénitaux du métabolisme des glucides actuellement identifiés: au cours des 5 dernières années nous avons pu identifier 17 cas de glycogénoses, 4 cas de galactosémie congénitale et 4 cas d'intolérance héréditaire au fructose.

Le diagnostic des glycogénoses ne pouvait jusqu'à maintenant être fait avec certitude que sur des prélèvements tissulaires (foie et muscle) dont la nature exclut la pratique courante. Nous avons étudié les possibilités de recherche et éventuellement d'identification de ces maladies sur des prélèvements sanguins.

Après un bref rappel des formes de glycogénoses actuellement identifiées, nous donnerons les résultats des dosages sanguins effectués dans un certain nombre des 17 cas de glycogénoses que nous avons identifiés: (a) dans les érythrocytes: dosage du glycogène; (b) dans les leucocytes: dosage du glycogène, de l'α-1,4-glucosidase et de la phosphorylase. Nous discuterons la valeur de ces tests dans la recherche des glycogénoses et nous comparerons nos résultats à ceux d'autres auteurs.

[1] Chargée de Recherche INSERM.

Classification des glycogénoses

On désigne sous le nom de glycogénose les états dans lesquels le contenu glycogénique des tissus est anormal quantitativement ou qualitativement. L'origine du trouble est une absence ou une anomalie d'une enzyme impliquée dans la glycogénèse ou dans la glycogénolyse. On connaît actuellement une dizaine de glycogénoses identifiées (tableau I). La fréquence de ces différents types est très variable: les glycogénoses de type III et de type I sont de loin les plus souvent rencontrées.

La répartition des 17 cas que nous avons étudiés s'établit de la manière suivante: 3 glycogénoses de type I; 1 glycogénose de type II; 8 glycogénoses de type III, dont 5 de type III A et 3 de type III B; 5 glycogénoses non identifiées. Ces résultats correspondent généralement dans leurs proportions aux fréquences relatives actuellement admises pour ces différentes formes.

Nous avons identifié ces glycogénoses par l'étude des tissus (foie, muscle, cœur) recueillis chez ces malades par prélèvement biopsique, chirurgical et, dans un cas, *post mortem*. Ces examens ont comporté: dosage du glycogène; étude de sa structure; mesure des activités enzymatiques impliquées dans la glycogénolyse: glucose-6-phosphatase, amylo-1,6-glucosidase, phosphorylase, maltase acide.

Nous nous sommes attachés, parallèlement, à étudier le sang de ces malades, et, quand cela était possible, celui de leurs parents. Nous rapportons d'abord les résultats des dosages effectués dans les érythrocytes puis dans les leucocytes.

Examen des érythrocytes

Le glycogène érythrocytaire a été dosé directement par le réactif à l'anthrone après extraction de l'hémolysat selon la technique de SIDBURY [5]. Les résultats sont exprimés en microgrammes de glycogène par gramme d'hémoglobine. Les contrôles ont été effectués chez des enfants normaux, dans les mêmes conditions de prélèvement (5 ml de sang veineux hépariné, recueilli le matin après 8 h de jeûne). Le taux moyen normal résultant de 20 dosages est de 222 μg/g d'hémoglobine.

1. Dans un cas de glycogénose de type I, nous avons trouvé un taux normal (220).

2. Dans 6 cas de limite-dextrinose, nous avons constaté des augmentations du glycogène érythrocytaire: ces augmentations ont été beaucoup

Tableau I. Classification des glycogénoses

Type	Principaux organes affectés	Déficit enzymatique	Taux de glycogène	Structure du glycogène
I Von Gierke	Foie – Rein	Glucose-6-phosphatase	Augmentation	Normale
II Pompe	Généralisée	α-1,4-Glucosidase acide	Augmentation	Normale
III. Limite-Dextrinose		Amylo-1,6-glucosidase:		
Sous-groupe A	Foie – Muscle	– absence totale	Augmentation	Limite dextrine
Sous-groupe B	Foie – Muscle	– absence partielle	Augmentation	Limite dextrine
IV. Andersen	Généralisée	Amylo-1,4—1,6-transglucosidase	Augmentation	Amylopectine
V. Mac Ardle	Muscle	Phosphorylase	Augmentation	Normale
VI. Hers	Foie	Phosphorylase hépatique	Augmentation	Normale
VII. Hers	Foie – Muscle	Phosphorylase-b-kinase	Augmentation	Normale
Lewis	Foie – Muscle	Glycogène synthétase	Diminution	Normale
Tarui	Muscle	Phosphofructokinase	Augmentation	Normale
Thompson	Muscle?	Phosphoglucomutase?	Augmentation	

plus importantes dans les 3 cas de sous-groupe B que dans les 3 cas de sous-groupe A. Une constatation semblable peut être faite sur les résultats publiés par VAN HOOF [7]: glycogénose de type III A (n = 10): moyenne 825 μg/g d'Hb (179 — 3.300); glycogénose de type III B (n = 7): moyenne 1068 μg/g d'Hb (358 — 7.885). Cet auteur [7] ne retient pas ces différences comme significatives; un nombre plus important de dosages serait nécessaire pour conclure à une augmentation plus nette du glycogène érythrocytaire dans les glycogénoses de type III B que dans le sous-groupe A, comme cela apparaît dans nos 6 cas.

3. Dans 2 cas de glycogénoses sans déficit enzymatique identifié, nous avons trouvé une augmentation modérée de glycogène érythrocytaire.

Tableau II. Glycogène érythrocytaire chez des sujets normaux, des malades et des parents de malades atteints de glycogénose (exprimé en microgrammes de glycogène par gramme d'hémoglobine)

Type		Nº	Taux de glycogène érythrocytaire
Témoins normaux			222±105 (N=20)
Glycogénoses			
I		15	220
III Sous-groupe A		10	442
		9	600
		13	797
Sous-groupe B		16	1 530
		6	2 058
		–	2 912
		14	1 170
Glycogénose non identifiée		8	499
		19	528
Parents			
II	Père	12	469
	Mère	–	553
III B	Père	14	165
	Mère	–	199

4. Nous avons étudié les parents de deux malades: dans un cas de maladie de Pompe, le père et la mère du malade présentaient une légère augmentation du glycogène érythrocytaire. Il est cependant hasardeux d'interpréter cet unique résultat dans le sens de la démonstration de l'hétérozygotisme des parents.

5. Enfin dans un cas de glycogénose de type III B, les parents du malade avaient une concentration de glycogène érythrocytaire normal.

A titre de comparaison nous donnons les résultats publiés par SIDBURY *et coll.* [6] (tableau III). L'étude de ces auteurs portait sur un cas de chaque catégorie de glycogénose. On pouvait constater une augmentation très importante du glycogène érythrocytaire dans la glycogénose de type III A, et une augmentation plus modérée dans un cas de type III B et dans un cas de type VI.

Nous avons également tenté d'étudier l'amylo-1,6-glucosidase chez des sujets normaux et chez des malades atteints de glycogénose de type III A selon la technique de HERS [8] mesurant l'incorporation du ^{14}C glucose au glycogène: de grandes variations dans les activités mesurées chez les témoins normaux et parmi les malades atteints de limite-dextrinose, ne nous ont pas permis d'interpréter ces résultats.

Tableau III. Glycogène érythrocytaire chez des sujets normaux et chez des malades atteints de Glycogénoses [1]

Type	No	Taux de glycogène érythrocytaire		Pourcentage de dégradation à la β-amylase
Témoins normaux	19	Sang du cordon	123	
	11	4 à 19 h	155	
	16	1 mois à 1 an	86	
	45	1 à 12 ans	70	44
	14	Adultes	56	43
I	22	1 à 15 ans	89	Normal
II	5	5 à 7 mois	99	Normal
III a	7	11 mois à 22 ans	3 507	22
III b	5	1 à 27 ans	410	20
IV	1	1 an	84	64
VI	10	10½ mois à 11 ans	341	42

[1] Selon SIDBURY et coll. [6].

Examen des leucocytes

Dans les leucocytes de témoins normaux et de malades atteints de diverses glycogénoses, nous avons dosé le glycogène, la maltase acide et la phosphorylase.

Dosage du glycogène leucocytaire

Dans les leucocytes isolés à partir de 5 à 6 ml de sang veineux hépariné, prélevé chez les sujets à jeun, nous avons pratiqué un dosage de glycogène suivant la méthode de CAROLL et coll. [1] (homogénéisation des leucocytes dans l'acide trichloracétique à 5 %, puis précipitation du glycogène par l'alcool à 99 % et dosage du précipité isolé par la méthode à l'anthrone) (Tableau IV.)

Tableau IV. Glycogène leucocytaire de parents de malades et de malades atteints de glycogénose (en microgrammes de glycogène par gramme de protéine)

Type		Nº	Glycogène leucocytaire
Sujets normaux			41±14 (N=22)
Glycogénoses:			
I		15	66
III Sous-groupe B		16	73
		14	40
Non identifiées		19	98
Parents:			
II	Père	12	48
	Mère	–	43
III B	Père	14	84
	Mère	–	68

1. Dans 22 dosages effectués chez des contrôles normaux (enfants de moins de 10 ans) nous trouvons un taux moyen de glycogène de 41 ± 14 µg/g de protéine.

2. Dans 1 cas de glycogénose de type I nous trouvons une augmentation significative (66).

3. Dans 2 cas de type III B nous trouvons un taux augmenté (73) et un taux normal (40).

4. Dans 1 cas de glycogénose non identifiée nous trouvons une augmentation importante (98).

5. Nous avons étudié les parents d'un enfant atteint de maladie de Pompe: le père et la mère avaient un taux de glycogène leucocytaire normal (48, 43).

6. Par contre le père et la mère d'un malade atteint de glycogénose de type III B ont présenté uue augmentation significative du glycogène leucocytaire (84 – 68).

Il semble donc, d'après ces résultats, que l'évaluation du taux de glycogène leucocytaire permette la recherche des glycogénoses; mais l'importance de l'augmentation est variable et ne permet pas une interprétation en faveur d'un type particulier de glycogénose. Dans le dépistage des porteurs hétérozygotes, nous n'avons étudié qu'un seul cas. Les résultats obtenus sont en faveur de ce test, mais nous manquons d'expérience pour en conclure avec certitude.

2. Etude de la maltase acide leucocytaire

Nous avons également dosé l'α-1,4-glucosidase leucocytaire (ou maltase acide de Hers) chez des enfants normaux de moins de 10 ans, chez un malade atteint de glycogénose généralisée du type II ou maladie de Pompe et chez les parents et frères et sœurs de ce même malade.

L'activité maltasique a été mesurée selon la technique de Illingworth [3]. Elle est exprimée en nanomoles de maltose hydrolysé par minute et par milligramme de protéines (tableau V).

Tableau V. α-1,4-Glucosidase leucocytaire de sujets normaux, des parents, des collatéraux et d'un malade atteint de glycogénose de type II (exprimée en nanomoles par minute et par milligramme de protéine)

		α-1,4-Glucosidase
Sujets normaux (enfants)		$5{,}7 \pm 2{,}3$ (N=20)
Malade		0,7
Parents:	Père	4,8
	Mère	6
Collatéraux	Frère	4,5
	Sœur C	3,4
	Sœur E	5,7
	Sœur O	5,2

1. Chez 20 sujets normaux (enfants de moins de 10 ans) nous avons trouvé une activité moyenne de 5,7 ± 2,3 unités.

2. Dans un cas de glycogénose généralisée de type II nous avons trouvé une activité maltasique presque nulle (0,7 unités).

3. Par contre les parents de ce malade et ses 4 frères et sœurs étudiés avaient tous des activités maltasiques leucocytaires normales.

Ce test semble donc valable pour dépister une glycogénose de type II, mais il ne peut être préconisé pour rechercher les porteurs de cette tare.

Les résultats obtenus par d'autres auteurs concordent avec les nôtres: HSIA [2] rapportant les résultats de NITOWSKY et GRÜNFELD [4], confirme l'absence de maltase acide leucocytaire chez les malades atteints de maladie de Pompe; par contre, à l'inverse de nos résultats, ils trouvent une activité enzymatique diminuée d'environ 50% chez l'un des parents de ce malade. Nous rapportons les résultats de ces auteurs (tableau VI).

Tableau VI. α-1,4-Glucosidase des leucocytes et des fibroblastes de parents de malades et de malade atteint de glycogénose de type II[1] (exprimée en micromoles par minute et par gramme de protéine)

	α-1,4-Glucosidase	
	Leucocytes	Fibroblastes
Témoins normaux	8±2,2	10,9±2,6
Malade		0
Parents	1,9	6,7
	4	4,5
	7	3,4

[1] D'après NITOWSKY et GRUNFELD. Reproduit par HSIA [2].

3. Autres activités enzymatiques leucocytaires

Outre le dosage de maltase acide, nous avons étudié la phosphorylase leucocytaire: nous utilisons la technique de SUTHERLAND rapportée par WALLIS et coll. [9] qui mesure cette activité en micromoles de glucose-1-phosphate hydrolysé par minute et par gramme de protéines. Chez 15 contrôles normaux (enfants de moins de 10 ans) nous trouvons une activité moyenne de 65,3 ± 10,6. Nous n'avons malheureusement pas pu étudier de glycogénose due à une absence ou à une anomalie de phosphorylase. Mais d'autres auteurs [10] ont constaté une activité phosphorylasique leucocytaire très diminuée (7,1 et 5,5 unités pour un taux normal de 29,6 ± 1,2) chez deux frères atteints de glycogénose de type VI.

Conclusions

Nos résultats nous amènent aux conclusions suivantes:

1. Le dosage du glycogène érythrocytaire peut être retenu pour rechercher certaines glycogénoses, mais il n'est pas valable pour toutes les catégories; il ne semble pas satisfaisant pour rechercher les porteurs de cette tare.
2. Les résultats des examens pratiqués dans les leucocytes semblent plus encourageants: le dosage du glycogène leucocytaire peut permettre de détecter une glycogénose mais non de l'identifier; il peut être utilisable pour rechercher les porteurs de cette tare mais nous manquons d'expérience pour en conclure.
3. Le dosage de maltase acide leucocytaire permet de détecter avec certitude la maladie de Pompe; mais ces résultats sont inconstants pour rechercher les porteurs de cette maladie.

Il semble donc que les dosages leucocytaires soient plus valables que les dosages érythrocytaires pour la recherche des glycogénoses; mais aucun ne semble absolument satisfaisant pour dépister les porteurs de cette tare.

Bibliographie

1. Caroll, N. V.; Longley, R. W. and Roe, J. H.: The determination of glycogen in liver and muscle by use of anthrone reagent. J. biol. Chem. *220:* 583 (1956).
2. Hsia, D. Y. Y.: The diagnosis and management of the glycogen storage disease. Amer. J. clin. Path. *50:* 44–51 (1968).
3. Illingworth Brown, B. and Zellweger, H.: α-1,4-Glucosidase activity in leucocytes from the family of two brothers who lack this enzyme in muscle. Biochem. J. *101:* 16 C (1966).
4. Nitowsky, H. M. and Grunfeld, H.: Lysosomal α-glucosidase in type II glycogenosis: activity in leukocytes and cell cultures in relation to genotype. J. Lab. clin. Med. *69:* 472–484 (1967).
5. Sidbury, J. B.; Cornblath, M.; Fischer, J. and House, E.: Glycogen in erythrocytes of patients with glycogen storage disease. Pediatrics *27:* 103 (1961).
6. Sidbury, J. B.; Gitzelmann, R. and Fischer, J.: The glycogenoses: further observations on glycogen in erythrocytes of patients with glycogenosis. Helv. paediat. Acta *16:* 506–516 (1961).

7. VAN HOOF, F.: Amylo-1,6-glucosidase activity and glycogen content of the erythrocytes of normal subjects patients with glycogen storage disease and heterozygotes. Europ. J. Biochem. *2:* 271–274 (1967).
8. VAN HOOF, F. and HERS, H. G.: The subgroups of type III glycogenosis. European J. Biochem. *2:* 265 (1967).
9. WALLIS, P. G.; SIDBURY, J. and HARRIS, R. C.: Hepatic phosphorylase defect. Amer. J. Dis. Child. *3:* 278–282 (1966).
10. WILLIAMS, H. E. and PROVINE, H. T.: Effect of oestradiol on glycogen synthetase in the rat uterus. Endocrinology *78:* 786 (1966).

Summary

In the present work the authors relate the studies they have performed in 17 cases of glycogenosis. They have identified these glycogenosis by studies of tissue samples (liver and muscles): determination of glycogen concentration and structure, measurement of the enzymatic activities of the glycogenolytic pathway. In blood sample of the same patients, and in some cases, of their parents, they have measured the average of glycogen content in red blood cells, in white blood cells and the following enzymatic activities: acid maltase and phosphorylase. In order to detect glycogenosis, tests performed in white blood cells seems to be more reliable than those performed in red blood cells. But none are quite reliable for the detection of carriers.

Adresse des auteurs: Dr M. MATHIEU et Dr J. COTTE, Laboratoire d'Enzymologie, Hôpital d'Enfants Debrousse, 29, rue Sœur-Bouvier, *F-69 Lyon 5e* (France)

7th int. Congr. clin. Chem., Geneva/Evian 1969; vol. 2: Clinical Enzymology, pp. 173–182
(Karger, Basel/München/Paris/New York 1970)

Elevation of Serum Enzyme Levels in Acute Poisoning

S. S. Brown, A. T. Proudfoot, J. A. Raeburn and N. Wright
University Department of Clinical Chemistry and Regional Poisoning Treatment Centre, Edinburgh

Introduction

Although the inhibition of serum or tissue enzymes in certain types of poisoning is well documented, there are only isolated reports of the elevation of serum enzymes. This paper reviews some findings in poisoned patients in an attempt to indicate the scope and limitations of this aspect of diagnostic enzymology.

In Britain, a total of about 70% of poisoning cases involve overdosage with hypnotic or analgesic drugs, or exposure to carbon monoxide [8]. Certain serum enzyme levels may be raised in these three types of poisoning and in the examples to be discussed (table I) it will be seen that the enzyme elevations may be due either to a specific toxic effect of the poison on a particular organ, or to a relatively non-specific effect, or to a combination of the two.

Patients and Methods

All patients were admitted to the Regional Centre in Edinburgh and treated according to the principles of intensive supportive therapy [7]. In no case was forced diuresis or haemodialysis used.

Plasma levels of paracetamol and of methaqualone were determined by direct spectrophotometry [9,6] and carboxyhaemoglobin by differential heat denaturation [14].

Serum alanine transaminase and aspartate transaminase were assayed by colorimetric "Auto Analyzer" techniques, hydroxybutyrate dehydrogenase and creatine kinase by recording spectrophotometry with 2-oxobutyrate and creatine phosphate as the respective substrates, and amylase by the starch-iodine method.

Table I. Elevation of serum enzyme levels in acute poisoning

Toxic substance		Enzyme elevated	Organ damage
Paracetamol (analgesic drug)	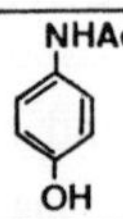	1. Alanine transaminase, GPT	Direct; specific
Carbon monoxide (town gas)	CO	2. Amylase	Direct; non-specific
Methaqualone (hypnotic drug)	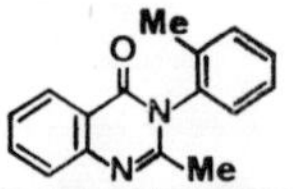	3. Asparate transaminase, GOT 4. Hydroxybutyrate dehydrogenase, HBD 5. Creatine kinase, CPK	? Indirect; ? non-specific

1 = L - alanine: 2-oxoglutarate aminotransferase
2 = α - 1,4 - glucan-4-gluconohydrolase
3 = L - aspartate: 2-oxoglutarate aminotransferase
4 = D - 2-hydroxybutyrate dehydrogenase
5 = ATP: creatine phosphotransferase

Results and Discussion

a) Paracetamol Poisoning

An apparently direct toxic effect on the liver follows overdosage with paracetamol, N-acetyl-p-aminophenol. It is not widely known that this is hepatotoxic, but the first fatal cases of poisoning, with acute centrilobular liver necrosis, were reported from Edinburgh in 1966 [4]. In our experience, about 30% of cases of paracetamol overdosage develop some measure of liver damage. In order to be able to assess the severity of this poisoning and to help in developing an effective treatment, we have reviewed liver function tests in a number of cases.

Table II shows the peak values of serum alanine transaminase, alkaline phosphatase, and bilirubin in seven patients. It can be seen that high values of transaminase may be reached, that there may be slight elevations of serum bilirubin, but that it is unusual to find raised alkaline phosphatase in such cases. The first patient, with a peak serum alkaline phosphatase of 30 King-Armstrong units %, was an example of fatal paracetamol poisoning; the other patients in this series made uneventful recoveries. The time course of the transaminase elevation is shown in figure 1. In the fatal case, the transaminase rose to one hundred times the normal level within 48 h of the inges-

Table II. Peak values of serum enzymes and of bilirubin in 7 cases of paracetamol poisoning

Patient	Sex	Age (year)	Alanine aminotransferase (SGPT; units/ml)	Alkaline phosphatase (K.A. units %)	Bilirubin (mg %)
1	F	18	4,000	30	4.2
2	F	40	2,000	11	2.0
3	M	26	2,000	13	5.3
4	M	39	1,500	9	1.0
5	F	31	500	8	0.5
6	F	25	60	10	0.9
7	F	16	46	9	0.4
Normal range			10–35	2–17	<1.0

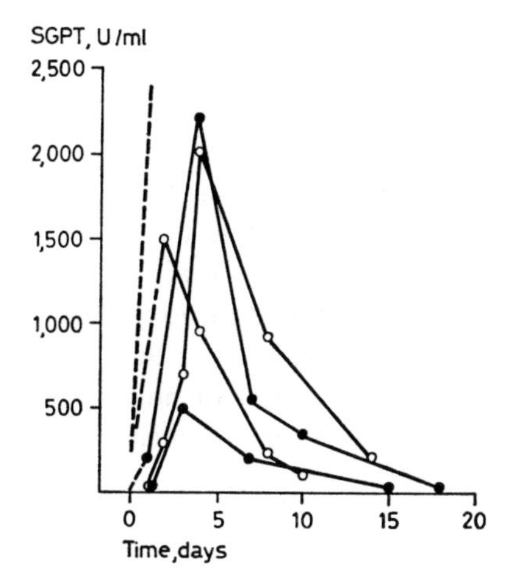

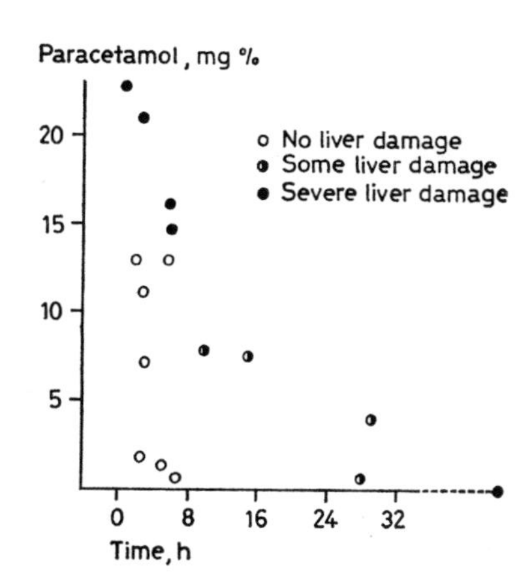

Fig. 1. Paracetamol poisoning in 5 patients: changes in serum alanine transaminase (GPT; units/ml) with time (days after ingestion).

Fig. 2. Paracetamol poisoning in 16 patients: occurrence of liver damage in relation to blood levels and to time after ingestion.

tion of an overdose, but in general peak values are reached within five days. In severe poisoning it may be several weeks before there is a return to normal.

The extent of liver damage in such cases cannot be related to the amount of drug absorbed (since in poisoned patients this is not known), but an attempt can be made to relate it to the blood level of N-acetyl-p-amino-phenol (fig. 2). The relationship is not a simple one. The problem is complicated by the fact that the peak level is reached within a few hours of ingestion but, in most cases, the drug is very rapidly removed from the circulation. In experimental subjects [12], up to 80% of a therapeutic dose is excreted in the urine, as conjugates of p-aminophenol, within 24 h. In the fatal case of paracetamol poisoning, there was no measurable blood level on admission to hospital 2 days after ingestion; she presented with severe hypoglycaemia and acidaemia and died within a few hours. We believe that any patient in whom unchanged paracetamol can be detected in the blood more than 24 h after ingestion is likely to develop some measure of liver damage, and that in the most severe poisoning this damage is overwhelming and irreversible. The biochemical picture is clearly one of hepatocellular damage with little obstructive element, and the toxic effect is primarily an organ-specific one. The present results show a striking similarity to the enzyme findings in experimental carbon tetrachloride poisoning [11].

b) Carbon Monoxide Poisoning

A *post mortem* finding of acute pancreatitis in a patient who died following exposure to town gas led us to examine the significance of serum amylase determinations in such patients. The results in 59 cases are shown in figure 3, with the peak serum amylase levels plotted against the blood carboxyhaemoglobin on admission. (Patients with carboxyhaemoglobin of less than 15% are not considered here, in order to exclude doubtful cases of poisoning.) In nearly all patients the peak serum enzyme levels were reached within one or two days of admission. Forty-nine of the patients were found to have serum amylase which did not exceed the upper limit of normal; a further six had levels of up to twice this value, and four patients had levels of up to ten times the normal. It is also evident that there is no correlation between the enzyme and the carboxyhaemoglobin levels–indeed the patients with the highest serum amylase levels all had carboxyhaemoglobin levels of less than 40%. This is perhaps not surprising since carboxyhaemoglobin

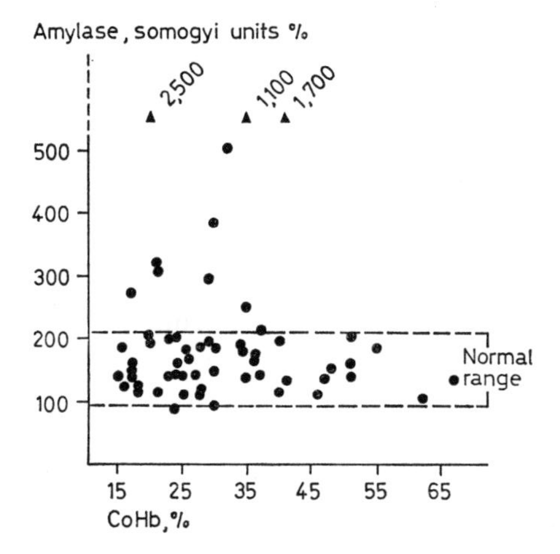

Fig. 3. Coal gas poisoning in 59 patients: plot of serum amylase and carboxyhaemoglobin levels.

per se is a poor measure of toxicity; the significance of a given level depends upon the total haemoglobin, the length of exposure to the lethal atmosphere, and the effectiveness of ventilation prior to admission to hospital.

Now, to the best of our knowledge, none of the standard works either on carbon monoxide poisoning [2] or on pancreatitis [13] suggests that there is a connection between these two conditions, although there is one report [5] of a case of pancreatic encephalopathy following town gas poisoning. What then is the explanation of these findings? We believe that a link is to be found in the well established myocardial toxicity which accompanies coal gas poisoning, and which is responsible of many deaths. In our experience, 30% of patients who are admitted with carboxyhaemoglobin levels of 15% or more have demonstrable myocardial ischaemia or develop myocardial infarction. In fact, two of these four patients with high serum amylase levels died after a few days with myocardial damage and raised aspartate transaminase and hydroxybutyrate dehydrogenase levels; the other two, non-fatal cases, had serum enzyme levels which were just at the upper limit of normal. We consider, therefore, that the elevation of serum amylase which may be found in carbon monoxide poisoning is a reflection of hypoxic injury to the pancreas, which is likely to be accompanied by myocardial (and also cerebral) damage. It is relevant that in myocardial infarctions generally, elevations of serum amylase have been reported [13], but that this apparently occurs in less than 5% of cases. Prolonged hypothermia and hypotension are also likely to play a part in precipitating this damage, but their precise role is difficult to determine.

c) Methaqualone Poisoning

In deeply unconscious patients suffering from overdosage with hypnotic or sedative drugs complex patterns of enzyme elevations may be found because there may be concurrent damage to the heart, kidney and liver. In the patients to be considered now, however, there was minimal renal or hepatic impairment. Our reason for studying such patients is that we wish to establish whether the cardiac arythmias which characterise severe overdosage with the tricyclic antidepressants [10] are produced by hypnotic drugs such as the barbiturates and methaqualone, and whether these drugs do have real cardiotoxic effects. We have, therefore, made serial determinations of the levels of aspartate transaminase, hydroxybutyrate dehydrogenase and creatine kinase in unconscious patients. The findings can best be illustrated by considering four particular cases of methaqualone poi-

Table III. Chief features of 4 cases of methaqualone poisoning

Patient	Sex	Age (year)	Plasma methaqualone (mg %)	Unconscious (Grade*/h)	Cardiac abnormalities	Clinical findings
1	F	66	4.8	III/60	None	Uneventful recovery
2	F	48	11.7	IV/45	None	Uneventful recovery
3	M	56	9.0	IV/85	History of myocardial infarction	Hypotensive; pneumonia; died (M.I.) on day 6
4	M	49	7.2	IV/100	Atrial extrasystoles	Hypotensive; hypothermic; pneumonia

* Depth of coma graded accorded to MATTHEW and LAWSON [1967].

soning (table III), but it must be emphasized that enzyme changes such as those which will be described appear to be produced by a wide variety of hypnotic drugs, and not just by methaqualone. Each of these patients was deeply unconscious for many hours, with quite high plasma methaqualone levels. The first two patients made an uneventful recovery with no clinical evidence of cardiac damage; patient three had a history of myocardial infarction, and died with a further infarction a few days after regaining consciousness; patient four showed minimal ECG changes, but without evidence of real myocardial damage.

The enzyme elevations which were found in these four patients are shown in figure 4, together with an indication of the pattern which typifies myocardial infarction. The feature which is common to all of these patients is the very steep rise in creatine kinase, but the time course of the rise is

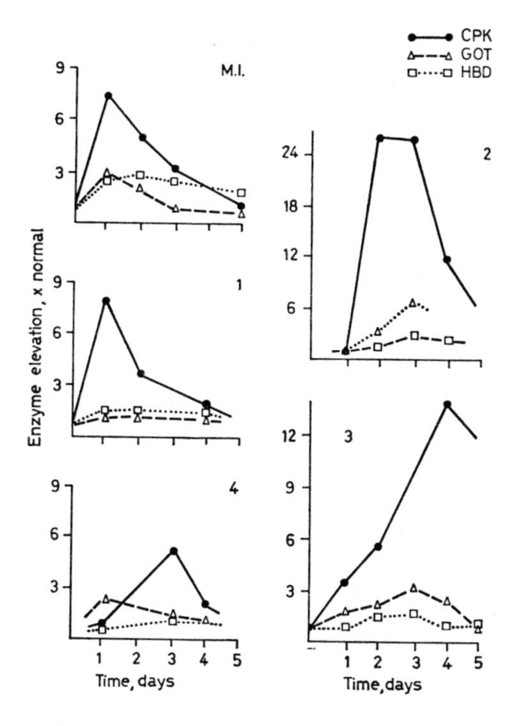

Fig. 4. Changes in serum enzymes (creatine kinase, CPK; aspartate transaminase, GOT; hydroxybutyrate dehydrogenase, HBD) in myocardial infarction and in 4 cases of methaqualone poisoning.

different in each case. The changes in aspartate transaminase and hydroxybutyrate dehydrogenase are less marked, and no consistent pattern is seen. Patient one showed no significant increase in either of these two enzymes but patient two, who also had no cardiac damage, showed a steep rise in aspartate transaminase after three days and a smaller rise in hydroxybutyrate dehydrogenase at about the same time. Patient three showed similar rises in these enzymes but both levels returned to normal before left ventricular failure and fatal myocardial infarction on the sixth day. Patient four, who had minimal ECG changes, showed an increase in aspartate transaminase which mimics that found in myocardial infarction but with no significant increase in hydroxybutyrate dehydrogenase.

It seems clear from these results that such enzyme elevations are not necessarily associated with cardiotoxicity. About 5% of unconscious poisoned patients do show ECG changes, with or without myocardial damage, but a higher proportion show enzyme changes of the type discussed here; these elevations are particularly pronounced in patients who are unconscious for a very long period of time. The most interesting and consistent finding in these cases is the very large increase in creatine kinase. Isoenzyme studies, by Dr. A. F. SMITH of the University Department of Clinical Chemistry in Edinburgh, show that this enzyme has its origin in skeletal muscle rather than heart or brain. We suggest, therefore, that generalised skeletal muscle damage accompanying coma is a prime cause of these enzyme elevations in poisoned patients. This effect is to be compared with the increase in creatine kinase which is associated with some forms of muscle disease or injury [3].

Conclusion

The three types of poisoning discussed here illustrate some of the serum enzyme elevations which may be found in straightforward cases of poisoning, where only one toxic substance is involved, and where there is a relatively uneventful clinical course. In more complex cases, there may be antecedent disease, injury or drug abuse, and concurrent or subsequent impairment of renal or pulmonary function with severe disturbance of acid-base status; bizarre patterns of enzyme elevations may then be seen whose significance is difficult to assess. Organ-specific damage may occur with some poisons, but the importance of non-specific effects due to hypoxia, hypotension and hypothermia [1], particularly in deeply unconscious patients, must not be underestimated.

Summary

In poisoned patients an increase in the level of a particular serum enzyme may reflect: a) a direct toxic effect of the poison on an organ from which the enzyme is derived: b) an indirect toxic action, i.e., secondary damage resulting from non-specific trauma; c) a combination of both direct and indirect effects. Clinical examples of these situations are discussed.

a) In 4 patients with paracetamol overdosage but showing few clinical signs of toxicity, serum alanine transaminase has been found to rise to 10 times the upper limit of normal, or more, within 2 to 4 days of ingestion of the overdose, with the return to normal levels delayed for up to 3 weeks; other liver function tests in these patients showed much less marked abnormalities. The toxic effect of the drug is therefore consistent with primary hepatocellular damage.

b) A survey of serum amylase levels in carbon monoxide-poisoned patients showed significant elevation of the enzyme level (up to 10 times the normal) in 4 out of 59 cases, 2 of which were fatal. This finding appears to reflect non-specific tissue damage following severe hypoxia.

c) Unconscious patients suffering from overdosage of hypnotic drugs may show a complex pattern of enzyme elevations. Thus, in 4 cases of severe methaqualone poisoning, relatively small increases in the levels of serum aspartate transaminase and hydroxybutyrate dehydrogenase were found, but with elevations of serum creatine kinase of up to 20 times the normal. Such changes seem to reflect generalized skeletal muscle damage, but more specific myocardial, renal, or hepatic involvement may also contribute to the overall picture.

References

1. Blyuger, A. F.; Belen'kii, M. L. and Shuster, Ya Ya.: Mechanism of increased serum enzyme activity during action of strong stimuli. Fed. Proc. *24:* Transl. Suppl. No 1, Pt. II, p. T 93 (1965).
2. Bour, H. and Ledingham, I. McA.: Carbon monoxide poisoning. Progr. Brain Res., vol. *24* (Elsevier, Amsterdam 1967).
3. Cohen, L.: Serum enzyme determinations: their reliability and value. Med. Clin. N. Amer. *53:* 115 (1969).
4. Davidson, D. G. B. and Eastham, W. N.: Acute liver necrosis following overdose of paracetamol. Brit. med. J. *ii:* 497 (1966).
5. Gaultier, M.; Fournier, E.; Gervais, P. et Bodin, I.: Encéphalopathie pancréatique survenue au décours d'une intoxication oxycarbonée: comparaison avec l'encéphalopathie post-intervallaire de l'intoxication oxycarbonée. Presse méd. *72:* 3263 (1964).
6. Lawson, A. A. H. and Brown, S. S.: Acute methaqualone (Mandrax) poisoning. Scott. med. J. *12:* 63 (1967).
7. Matthew, H. and Lawson, A. A. H.: Treatment of common acute poisonings (Livingstone, Edinburgh 1967).
8. Matthew, H.; Proudfoot, A. T.; Brown, S. S. and Aitken, R. C. B.: Acute poisoning: organization and work-load of a treatment centre. Brit. med. J. *iii:* 489 (1969).

9. ROUTH, J. I.; SHANE, N. A.; ARREDONDO, E. G. and PAUL, W. D.: Determination of N-acetyl-p-aminophenol in plasma. Clin. Chem. *14:* 882 (1968).
10. STEEL, C. M.; O'DUFFY, J. and BROWN, S. S.: Clinical effects and treatment of imipramine and amitriptyline poisoning in children. Brit. med. J. *iii:* 663 (1967).
11. VOLEK, V.: Acute hepatitis. Acta Univ. Carol. Med., Monogr. *31:* 43 (1967).
12. WELCH, R. M. and CONNEY, A. H.: A simple method for quantitative determination of N-acetyl-p-aminophenol (APAP) in urine. Clin. Chem. *11:* 1064 (1965).
13. WHITE, T. T.: Pancreatitis (Arnold, London 1966).
14. WHITEHEAD, T. P. and WORTHINGTON, S.: The determination of carboxyhaemoglobin. Clin. chim. Acta *6:* 356 (1961).

Authors' address: Dr. S. S. BROWN, Dr. A. T. PROUDFOOT, Dr. J. A. RAEBURN and Dr. N. WRIGHT, University Department of Clinical Chemistry and Regional Poisoning Treatment Centre, *Edinburgh* (Scotland)

7th int. Congr. clin. Chem., Geneva/Evian 1969; vol. 2: Clinical Enzymology, pp. 183–186
(Karger, Basel/München/Paris/New York 1970)

Influences on Serum Monoamine Oxidase (MAO) and Diamine Oxidase (DAO) Activity

N. TRYDING, G. TUFVESSON and S. E. NILSSON
Department of Clinical Chemistry and Internal Medicine, Central Hospital Kristianstad

Human blood serum contains at least two different amine oxidases that catalyze the oxidative deamination of amines. One has monoamines as the substrate of choice and is called MAO, the other prefers diamines and is called DAO.

The soluble serum MAO catalyzes the oxidation of monoamines, e.g., benzylamine, butylamine, dopamine, kynuramine, tryptamine and tyramine [1]. However, some other biologically important monoamines such as adrenalin, noradrenalin and serotonin are not attacked by human serum MAO. In the measurement of serum MAO benzylamine has proved a suitable substrate [2]. We have also used ^{14}C-labelled substrates, such as benzylamine, tyramine [3] and butylamine.

The serum-MAO level is sometimes altered in the presence of internal diseases and during medicinal treatment. We have confirmed the findings of MCEWEN *et al.* [4, 11] that serum MAO is increased in chronic congestive heart failure, in which the increase varies with the degree of incompensation, in liver cirrhosis, and in thyrotoxicosis. Recently, we also found a significant increase of serum MAO concentration in persons with diabetes mellitus [5]. The enzyme activity is increased already from the very onset of the diabetic symptoms and does not vary with the type or duration of therapy. In first-degree relatives of patients with diabetes mellitus, however, the serum-MAO level proved normal [6]. Nor was the serum enzyme level elevated in persons with borderline results of glucose tolerance tests [6]. The serum-MAO values do not seem to be elevated until the disease has reached the hyperglycemic state. It should be mentioned that some MAO inhibitors potentiate the effects of insulin and counteract diabetes mellitus [7, 8]. Diabetics with chronic congestive heart failure or thyrotoxicoses have especially high MAO values. A decreased serum-MAO activity was sometimes found in some cases of malignant lymphomas and during chronic administration of glucocorticosteroids [5].

The serum-MAO level in 25 women (mean age: 27 years) was studied before and after 6 months of regular use of contraceptive steroids (3 mg norethisteron acetate and 0.05 mg etinylestradiol a day on 21 consecutive days each month). As shown in table I the MAO activity was found to be significantly decreased. Strangely enough, the serum-MAO level is normal in pregnancy [9, 16].

We have studied the influence of age and sex on the mean serum-MAO activity [6]. The level was highest in prepuberal children. In puberty the enzyme activity decreased abruptly to adult level, at which it persisted in females up to 50 years of age and in males up to 65, when it again rose, though less markedly. In the presence of diabetes mellitus the MAO level was about 50% higher, at the upper 2 S.D. border of the normal range, but the correlation with age was still the same as in the controls [6].

The mechanism of the endocrine regulation of serum-MAO activity is unknown. Our findings may suggest that the level of the serum-MAO activity is dependent on sex hormones.

We have not observed any correlation between serum-MAO and liver-function tests or blood-coagulation activity in normal persons [10]. It is, however, possible that diabetics with elevated MAO levels may have liver disturbances not detected by conventional liver-function tests. As pointed out by MCEWEN the liver may play a central role in the regulation of serum-MAO activity [11].

Diamine oxidase (DAO) catalyzes the oxidative deamination of histamine and diamines, e.g., putrescine and cadaverine [12]. For routine measurement of serum DAO, we use ^{14}C-putrescine as the substrate [13, 14].

Table I. Serum MAO (mU/1) and DAO (mU/1) in 20 women before and after 6 months of regular use of contraceptive steroids*

	MAO		DAO	
	Mean	S. D.	Mean	S. D.
Before use	475	93	5.1	3.9
After 6 months	383	87	3.7	2.5

* ELMFORS, B.; TRYDING, N. and TUFVESSON, G.

In human serum the DAO activity is normally low, around 4 mU/l [13]. Among apparently healthy persons a few have a distinctly higher enzyme level than the majority of the group. The DAO values were significantly lower in children than in adults [13]. In contrast with the activity of MAO that of DAO is not increased in diabetes. Preliminary results [15] show that serum DAO is occasionally increased in patients with duodenal ulcer but not in those with gastric ulcer. DAO is not significantly altered by contraceptive steroids (table I). The activity is increased about 500-fold during the last trimester of normal pregnancy [13]. The increase derives from the placenta. Determination of serum-DAO activity is used in clinical practice in gynecological diseases and obstetrical disorders, e.g., threatened abortion and mola hydatidosa (chorionepithelioma).

SOUTHREN *et al.* have shown that in pregnancy complicated by diabetes mellitus the DAO activity is often abnormally low [16]. In these cases the fetuses are thought to be exposed to a continuous high risk. We have not found such low DAO values as reported, but we have confirmed that serum DAO is lower in pregnancy complicated by diabetes than in normal pregnancy [17].

We have shown, that parenteral administration of heparin produces a prompt and marked increase of the DAO activity in human blood serum [18, 19]. Intravenous administration of heparin in therapeutic doses results in an increase of the DAO activity to levels otherwise seen only during the latter part of pregnancy. This effect of heparin varies with the route of administration and dose given, and appears to be ascribable to a release of DAO into the blood stream and lymph. After intravenous injection of heparin into pregnant women the DAO activity is increased to extremely high values [20]. HANSSON has recently shown that postheparin DAO can be separated from pregnancy DAO by electrophoresis and that these two enzyme forms are eliminated from the blood at fundamentally different rates [21]. The half-life is a few minutes for postheparin DAO and about 20 h for pregnancy DAO.

References

1. MCEWEN, C. M.: Human plasma monoamine oxidase. J. biol. Chem. *240:* 2003 (1965).
2. MCEWEN, C. M. and COHEN, J. O.: An amine oxidase in normal human serum. J. Lab. clin. Med. *62:* 766 (1963).
3. TUFVESSON, G.: Determination of monoamine oxidase activity in human blood serum with ^{14}C-labelled substrates. Scand. J. clin. Lab. Invest. *23:* 71 (1969).

4. McEWEN, C. M. and HARRISON, D. C.: Abnormalities of serum monoamine oxidase in chronic congestive heart failure. J. Lab. clin. Med. *65:* 546 (1965).
5. NILSSON, S. E.; TRYDING, N. and TUFVESSON, G.: Serum monoamine oxidase (MAO) in diabetes mellitus and some other internal diseases. Acta. med. scand. *184:* 105 (1968).
6. TRYDING, N.; NILSSON, S. E.; TUFVESSON, G.; BERG, R.; CARLSTRÖM, S.; ELMFORS, B. and NILSSON, J. E.: Physiological and pathological influences on serum monoamine oxidase. Scand. J. clin. Lab. Invest. *23:* 79 (1969).
7. COOPER, A. J. and ASHROFT, G.: Modification of insulin and sulfonylurea hypoglycemia by monoamine oxidase inhibitor drugs. Diabetes *16:* 272 (1967).
8. WICKSTRÖM, L. and PETTERSSON, K.: Treatment of diabetics with monoamine oxidase inhibitors. Lancet *ii:* 995 (1964).
9. McEWEN, C. M.: Serum amine oxidase in pregnancy. J. Lab. clin. Med. *64:* 540 (1964).
10. TRYDING, N.; BERG, R.; NILSSON, J.-E.; NILSSON, S. E.; STRANDLIE, O. K. and TUFVESSON, G.: Determination of coagulation activity in the normal and near normal area. Farmakoterapi *25:* 27 (1969).
11. McEWEN, C. M. and CASTELL, D. O.: Abnormalities of serum monoamine oxidase in chronic liver disease. J. Lab. clin. Med. *70:* 36 (1967).
12. ZELLER., E. A.: Identity of histaminase and diamine oxidase. Fed. Proc. *24:* 766 (1965).
13. TUFVESSON, G. and TRYDING, N.: Determination of diamine oxidase activity in normal human blood serum. Scand. J. clin. Lab. Invest. *24:* 163 (1969).
14. TRYDING, N. and WILLERT, B.: Determination of plasma diamine oxidase (histaminase) in clinical practice. A comparison between a biological method and radiochemical micromethod. Scand. J. clin. Lab. Invest. *22:* 29 (1968).
15. TRYDING, N.; HANSSON, R.; JONSON, G.; NILSSON, S. E.; SKÖLLING, R. and TUFVESSON, G.: (to be published).
16. SOUTHREN, A. L.; WEINGOLD, A. B.; KOBAYASHI, T.; SHERMAN, D. H.; GRIMALDI, R. and GOLD, E. M.: Diamine oxidase in diabetic pregnancy. Amer. J. Obstet. Gynec. *101:* 899 (1968).
17. PERSSON, B.; TRYDING, N. and TUFVESSON, G.: (to be published).
18. TRYDING, N.: Heparin-induced diamine oxidase (DAO) activity. Scand. J. clin. Lab. Invest. *17:* Suppl. 86: 196 (1965).
19. DAHLBÄCK, O.; HANSSON, R.; TIBBLING, G. and TRYDING, N.: The effect of heparin on diamine oxidase and lipoprotein lipase in human lymph and blood plasma. Scand. J. clin. Lab. Invest. *21:* 17 (1968).
20. HANSSON, R.; TRYDING, N. and TÖRNQVIST, Å.: Diamine oxidase (histaminase) in human pregnancy. Acta obstet. gynec. *48:* 8 (1969).
21. HANSSON, R.: Diamine oxidase isoenzymes in human blood plasma. Scand. J. clin. Lab. Invest. *25:* 33 (1970).

Authors' address: Dr. N. TRYDING, Dr. G. TUFVESSON and Dr. S. E. NILSSON, Department of Clinical Chemistry, Central Hospital Kristianstad, S29100 *Kristianstad* (Sweden)

7th int. Congr. clin. Chem., Geneva/Evian 1969; vol. 2: Clinical Enzymology, pp. 187–194
(Karger, Basel/München/Paris/New York 1970)

Activités tréhalasiques du sérum et de l'urine chez l'homme

J. F. DEMELIER, C. BARK, J. LABAT et J. E. COURTOIS
Laboratoires de Biochimie de l'Hôpital Laennec et de la Faculté de Pharmacie de Paris [1]

I. Généralités

La tréhalase α, α'-glucoside-1-glucohydrolase (E.C.3.2.1.28) scinde le tréhalose, disaccharide non réducteur en deux molécules de glucose. Le dosage de cet hexose permet d'apprécier l'activité tréhalasique [2, 9]. Les résultats pour le sérum sont exprimés en Unités Tréhalasiques (U. Tr.); l'U. Tr. correspond au nombre de μM de tréhalose hydrolysées par 1 et par min à pH 5,2 et à 37° C.

II. Etude de quelques caractéristiques de l'activité tréhalasique sérique

Une importante activité tréhalasique a été constamment mise en évidence dans le sérum de sujets sains [9]. L'activité moyenne d'une centaine de déterminations était de 14,2 U. Tr. avec un écart type de 9,40. Il existe de nettes variations de la tréhalasémie chez les sujets normaux. Les valeurs groupées par âge s'étagent de 1,1 à 56,5 U. Tr. [6]. Il n'y a pas de variations significatives chez l'adulte en fonction de l'âge. D'une façon régulière, l'activité moyenne des sérums d'hommes est supérieure à celle des sérums de femmes du même groupe d'âge.

Chez l'enfant [1], l'activité tréhalasique augmente régulièrement avec l'âge depuis la naissance pour atteindre le taux de l'adulte entre 6 et 11 ans.

[1] E.R.A. n° 99, Osides et Osidases du C.N.R.S.

III. Etude de quelques caractéristiques de l'activité tréhalasique urinaire

Nous avons évalué l'activité tréhalasique des urines selon le protocole suivant: à 1 ml de solution M/20 de tréhalose nous ajoutons: 0,5 ml de tampon acide acétique M/1 – NaOH de pH 5,7, 1 ml de solution de NaF, 1 ml d'urine fraîchement collectée et de l'eau distillée pour compléter le volume à 5 ml.

Après 6 h d'incubation à 37° C, le glucose libéré est déterminé selon la microméthode au réactif cuproalcalin de Nelson et Somogyi. Un essai témoin sans tréhalose est réalisé parallèlement pour permettre de soustraire le pouvoir réducteur de l'urine; ce dernier s'accroît légèrement après l'émission. L'activité est exprimée en U. Tr.: μ mol de substrat hydrolysées par min pour 1 l d'urine à pH 5,7 et 37° C.

L'urine humaine renferme régulièrement une faible activité tréhalasique. Les valeurs obtenues sur des échantillons prélevés à différentes heures s'échelonnent de 0,44 à 1,8 U. Tr. [4]; la moyenne de ces déterminations est de 0,93 U. Tr. pour 23 déterminations chez l'homme et 0,88 U. Tr. pour 21 déterminations chez la femme; il existe comme pour la tréhalase sérique de nettes variations de l'activité entre différents sujets, et chez un même sujet, lors des différents prélèvements.

L'élimination urinaire de l'enzyme est indépendante de la tréhalasémie; la mesure de la clearance conduit à des valeurs s'échelonnant de 0,01 à 0,60. D'autre part il n'y a aucun rapport entre l'activité tréhalasique urinaire et la teneur de l'urine en chlorures, urée et créatinine; le rapport U. Tr./créatinine (en g/l) a oscillé entre 0,17 et 9,0 pour des échantillons d'urines de 24 h; il s'est situé entre 0,15 et 7,7 pour diverses fractions émises au cours de la journée. Ces résultats montrent que l'enzyme urinaire ne provient pas d'une simple filtration glomérulaire de l'enzyme sérique.

IV. Etude de quelques propriétés de la cinétique des tréhalases sérique et urinaire

Ces disaccharidases sont assez thermostables puisqu'elles conservent toute leur activité après un chauffage de 10 min à 50° C et ne sont totalement inactivées que vers 60° C.

Les résultats groupés dans le tableau I montrent que le pH optimum de l'enzyme sérique et celui de l'enzyme urinaire sont très proches. Il en est de même de leurs constantes de Michaelis déterminées au pH optimum et à 37° C.

Tableau I. pH optimum et constante de Michaelis de quelques tréhalases humaines

Origine	Rein	Sérum	Urine
pH	5,7 – 5,8	5,7	5,6 – 5,8
K_M	3×10^{-3}	1,8 à 3,0 $\times 10^{-3}$	2,3 à 6,0 $\times 10^{-3}$

Ces valeurs sont du même ordre de grandeur que celles obtenues avec la tréhalase rénale humaine [7].

V. Action de différents effecteurs

Au cours de travaux antérieurs [5, 7] nous avons constaté que les tréhalases rénale et sérique étaient fortement inhibées par la phloridzine. Cet hétéroside, β-glucoside d'une dihydrochalcone, est connu pour provoquer une glycosurie par absence de réabsorption tubulaire. La localisation d'une grande partie de la tréhalase dans les tubules rénaux [11] confère une grande importance à cette inhibition.

Nous avons étudié plus en détail le comportement des tréhalases sérique et urinaire en présence de phloridzine (I), de son aglycone le phlorétol (II) et du naringoside (III) dont la structure est proche de celle des dihydrochalcones, la différence portant sur une cyclisation en flavanone.

Nos résultats sont groupés dans le tableau II où nous les avons comparés à ceux que nous avons obtenus au cours d'un travail effectué parallèlement sur les tréhalases du rein d'Homme, du rein et de l'intestin de Porc [5, 7, 10].

Les enzymes sérique et urinaire sont inhibées par la phloridzine; cette inhibition est dans les deux cas typiquement compétitive; nous avons d'ailleurs obtenu le même type d'inhibition avec l'enzyme rénale de l'Homme.

Le phlorétol provoque une inhibition du même type mais beaucoup plus intense compte tenu de sa faible solubilité. Il faut noter que cette inhibition n'a pu être mise en évidence avec la tréhalase sérique qu'après purification partielle de cette enzyme par électrophorèse préparative couplée à une élution continue des fractions séparées. Nous ne conservons que les tubes correspondant aux α-globulines du sérum qui contiennent toute l'activité tréhalasique. L'absence d'inhibition avant purification est vraisemblablement due

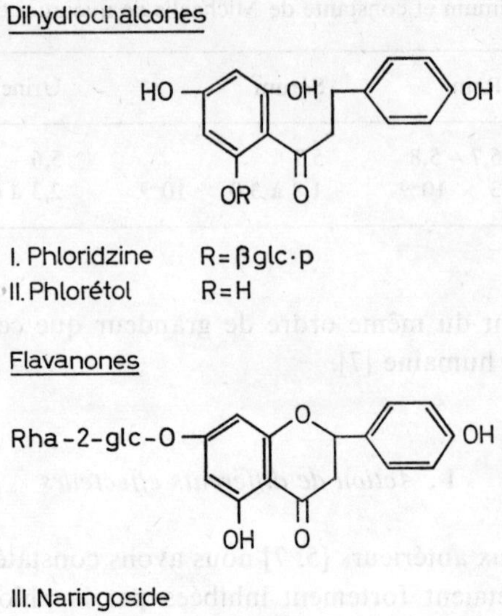

Dihydrochalcones

I. Phloridzine R = β glc·p

II. Phlorétol R = H

Flavanones

III. Naringoside

Fig. 1. Structure des dérivés de dihydrochalcones et de flavanones étudiés.

Tableau II. Action de différents effecteurs sur des tréhalases de mammifères de diverses origines

		Homme			Porc	
		Rein	Sérum	Urine	Rein	Intestin
Phloridizine	0,01 M	43*	41	56	79	57
Phlorétol	0,00014 M	25	56	83	41	25
Naringoside	0,01 M	0	0	0	0	0
Saccharose	0,01 M	48	30	63	40	45

* Les chiffres du tableau indiquent les pourcentages d'inhibition à pH 5,7 et 37°C.

à l'adsorption du phlorétol sur les protéines sériques que nous avons éliminées.

Nous n'avons obtenu aucune inhibition des tréhalases sérique et urinaire par le naringoside; cet hétéroside n'a pas inhibé non plus les tréhalases rénales et intestinales [5, 7, 10].

Toutes les tréhalases de mammifères étudiées ont été inhibées compétitivement par le saccharose. Cette inhibition est vraisemblablement due à l'analogie de structure de ce disaccharide avec le tréhalose.

VI. Variations de la tréhalase sérique et urinaire dans quelques cas pathologiques

Lors de travaux antérieurs, nous avons localisé la tréhalase chez l'Homme [3]. Nous n'avons trouvé cette enzyme que dans trois organes: rein, intestin grêle et foie. Le rein possède de loin la plus forte activité; dans le foie, l'activité est par contre minime. Nous avons purifié la tréhalase rénale humaine et étudié ses propriétés [7]. Nous avons ensuite effectué une première purification de l'enzyme intestinale de Porc [10]. Ces travaux sont poursuivis parallèlement à l'étude des tréhalases sérique et urinaire dans le but de comparer leurs propriétés pour déterminer leur origine et leur rôle.

L'absence de corrélation entre le taux de la tréhalase sérique et le débit urinaire de l'enzyme permet de penser que la faible activité tréhalasique retrouvée dans l'urine ne provient pas d'une excrétion rénale de la tréhalase du sang. Il semble plus vraisemblable qu'elle soit en rapport avec une élimination de l'enzyme provenant des cellules tubulaires rénales où paraît se localiser la tréhalase [SACKTOR, 11].

Deux organes sont susceptibles de laisser diffuser de la tréhalase dans le sérum: le rein et le foie. Nous avons pu étudier l'activité tréhalasique de six sujets binéphrectomisés [8]. L'ablation des deux reins datait pour certains de plusieurs mois ou même d'un an (tableau III).

Chez tous ces malades, il persiste une nette activité tréhalasique sérique. Le taux moyen de cette activité est de 8,3 U. Tr. Il est un peu inférieur à celui des sujets sains. Il est donc possible que le rein participe pour une faible part à la diffusion de la tréhalase dans le sang; cependant, la persistance d'une nette activité tréhalasique dans le sérum laisse penser que la tréhalase sérique a une autre origine.

Le foie semble être l'organe qui préside à la synthèse de la plus grande partie de l'enzyme sérique. Nous avons pu observer un abaissement net et assez constant de la tréhalasémie dans les cirrhoses décompensées avec atteinte cellulaire sévère. Ceci semble être en faveur de notre interprétation. Nous avons d'autre part régulièrement retrouvé une forte augmentation de la tréhalasémie chez les sujets atteints d'obstruction des voies biliaires principales et présentant un syndrome rétentionnel. La bile humaine ren-

Tableau III. Activité tréhalasique de sérums de sujets binéphrectomisés

Sujet	1	2	3	4	5	6
Sexe	♀	♂	♀	♂	♂	♀
Age (en années)	28	22	16	33	24	35
Intervalle entre l'intervention et le prélèvement de sang (en jours)	0 90	30	240 360	17 24	0 23	270 277
Activité (en U.Tr.)	8,7 16,2	5,4	15 14	5 10	3 1,5	3 5

ferme de la tréhalase [3] et semble être une voie d'élimination de cette enzyme.

Outre ces hypertréhalasémies mises en évidence dans les cas d'obstruction biliaire, un chercheur de notre laboratoire a découvert en 1962 une forte augmentation de la tréhalase sérique chez une diabétique d'origine martiniquaise [9]. Nous avons pu obtenir de nouveaux échantillons de sang de ce sujet et montrer que sa tréhalasémie oscillait aux environs de 80 U. Tr.; elle se maintient ainsi à un taux constamment élevé depuis 7 ans (tableau IV).

Ce taux est indépendant de la glycémie et le diabète ne paraît présenter aucune autre caractéristique particulière. Cette hypertréhalasémie n'a d'ailleurs jamais pu être retrouvée chez d'autres sujets d'origine antillaise ou chez d'autres diabétiques.

VII. Discussion

Il ne nous est pas encore possible d'attribuer un rôle à la tréhalase chez l'Homme. Le rein et l'intestin sont les deux organes les plus riches en tréhalase. Ce sont aussi les deux seuls susceptibles d'accumuler le glucose contre un gradient de concentration. Il est possible que la tréhalase intervienne dans le transport actif du glucose à ces niveaux. Dès 1962, COURTOIS *et al.* [2] avaient émis l'hypothèse que la tréhalase servait à remettre en circulation le glucose condensé en tréhalose; la formation de tréhalose-phosphate par transfert du groupement α-D-glucopyranosyl activé dans l'UDP-glucose sur le carbone 1 du glucose-6-phosphate n'étant pas incompatible avec les processus habituels du métabolisme intermédiaire. Le tréhalose-phosphate

Tableau IV. Teneur en tréhalase sérique d'un diabétique hypertréhalasémique

Date de l'analyse	Glycémie en g/l	Activités tréhalasiques
23. 2. 1962	1,75	85
27. 3. 1962	2,90	86
27. 9. 1962	2,80	58
2. 4. 1963	2,20	83
26.11. 1964	1,50	53
12. 5. 1965	2,10	75
7.12. 1967	1,10	76
25. 9. 1969	1,15	95

serait alors hydrolysé par une phosphatase avec libération de tréhalose. Cette hypothèse, reprise depuis par SACKTOR [11], n'avait jusqu'à présent pas été confirmée. En effet, le tréhalose n'avait jamais été mis en évidence chez les mammifères. VAN HANDEL [13] en particulier, injectant du glucose marqué à des lapins et à des rats n'avait pas réussi à mettre en évidence du tréhalose marqué dans les reins de ces animaux. Très récemment, cependant, SACKTOR et BERGER [12] incubant pendant un temps très court des tranches de rein de lapin en présence de glucose marqué ont réussi à observer la biosynthèse de tréhalose marqué. Cette découverte, étayant l'hypothèse précédemment émise, devrait permettre de trouver une explication à la présence de tréhalase dans le sérum et aider ainsi à mettre en évidence d'éventuelles variations pathologiques du taux sérique ou urinaire de cette enzyme.

Résumé

Le sérum et l'urine humaine contiennent régulièrement de la tréhalase. Les propriétés de ces enzymes sont très voisines et similaires à celles de la tréhalase rénale de l'homme. La phloridzine et le saccharose sont des inhibiteurs compétitifs. La tréhalase du sérum paraît provenir pour une partie du rein et pour une autre du foie. La tréhalasémie est nettement abaissée chez les cirrhotiques, elle augmente dans les ictères par obstruction. Le taux est faiblement abaissé chez les sujets binéphrectomisés. La tréhalase urinaire a un taux indépendant de la tréhalase sérique; elle semble provenir des tubules rénaux.

Summary

Human serum and urine always exhibit a trehalase activity. The properties of these enzymes are very similar and closely resembling those of renal human trehalase. Phlorizin and sucrose are competitive inhibitors. Serum trehalase seems to be produced partly by the kidney, partly by the liver. Serum trehalase decreases significantly in cirrhose; it increases in obstructive jaundice. Its level decreases slightly in binephrectomy. Urine trehalase which varies independantly from serum trehalase seems to be produced by renal tubules.

Bibliographie

1. COURTOIS, J. E.; CHERON, J. M. et DEMELIER, J. F.: Influence de l'âge sur l'activité tréhalasique du sérum chez l'homme et le lapin. Bull. Soc. Chim. biol. *48:* 1279–1280 (1966).
2. COURTOIS, J. E.; DEBRIS, M. M. et GEORGET, J. Cl.: Mise en évidence d'une tréhalase dans le sérum humain. Bull. Acad. nat. Méd. *146:* 599–602 (1962).
3. COURTOIS, J. E. et DEMELIER, J. F.: Répartition de la tréhalase chez l'homme et quelques mammifères. Bull. Soc. Chim. biol. *48:* 277–286 (1966).
4. COURTOIS, J. E.; DEMELIER, J. F.; LABAT, J. et BARCK, C.: Recherches sur la tréhalase urinaire humaine. C. R. Soc. Biol. *10:* 2094–2097 (1969).
5. COURTOIS, J. E.; DEMELIER, J. F.; LABAT, J. et HARGREAVES, F.: Séparation et propriétés de la tréhalase rénale du porc. Bull. Soc. Chim. biol. *50:* 1291–1304 (1968).
6. COURTOIS, J. E.; GEORGET, J. Cl. et DEMELIER, J. F.: Activité tréhalasique du sérum humain chez les adultes. C. R. Soc. Biol. *160:* 534–536 (1966).
7. DEMELIER, J. F. et COURTOIS, J. E.: Séparation et propriétés de la tréhalase rénale humaine. Bull. Soc. Chim. biol. *50:* 2553–2560 (1968).
8. DEMELIER, J. F.; SACHS, C. et COURTOIS, J. E.: La tréhalase du sérum chez les sujets binéphrectomisés. C. R. Soc. Biol. *162:* 1924–1925 (1968).
9. GEORGET, J. Cl.: Recherches sur la tréhalase du sérum humain. Th. Doct. univ. Pharm. Paris (1967).
10. LABAT, J.; DEMELIER, J. F., and COURTOIS, J. E.: Comparison of human and hog renal trehalases and hog intestinal trehalase. Abstr. 6th Meeting F.E.B.S. Madrid, p. 288 (1969).
11. SACKTOR, B.: Trehalase and the transport of glucose in the mammalian kidney and intestine. Proceed. nat. Acad. Sci. US. *60:* 1007–1014 (1968).
12. SACKTOR, B. and BERGER, S. J.: Formation of trehalase from glucose in the renal cortex. Biochem. biophys. Res. Comm. *6:* 796–800 (1969).
13. VAN HANDEL, E.: Do trehalose and trehalase function in renal glucose transport? Science *163:* 1075–1076 (1969).

Adresse de l'auteur: J. F. DEMELIER, Laboratoire de Biochimie, Hôpital Laennec, 42, rue de Sèvres, *75-Paris* 7e (France)

7th int. Congr. clin. Chem., Geneva/Evian 1969; vol. 2: Clinical Enzymology, pp. 195–202 (Karger, Basel/München/Paris/New York 1970).

Cholinestérase et arylestérase sériques dans le diabète

D. VINCENT, M. PLAUCHU et M. GIRARD

Laboratoire de Biochimie appliquée (Prof. D. VINCENT), Service de Médecine et Centre du Diabète, Hôtel-Dieu (Prof. M. PLAUCHU) et Clinique médicale D, Hôpital de la Croix-Rousse (Prof. M. GIRARD). Faculté de Médecine, Lyon

L'augmentation d'activité de la cholinestérase sérique (pseudocholinestérase) dans le diabète est un fait depuis longtemps et maintes fois signalé [DROUET *et al.*, 6; M. FABER, 8; DE PRAT, 14; KASWIN, 11; CRISTOL *et al.*, 4; DHELLEMMES 5; VINCENT et PARANT, 18; DUCARRE, 7, etc.]. Mais cette constatation n'a fait l'objet que de recherches partielles sur des cas peu nombreux et sans référence aux diverses variétés de ce syndrome métabolique. Aussi nous a-t-il paru intéressant d'étendre cette étude à un plus grand nombre de cas bien définis, et aussi à une autre estérase carboxylique, plus récemment découverte, l'arylestérase [ALDRIGE, 1; MOUNTER et WHITTAKER, 12] ou estérase A d'AUGUSTINSSON [2], non encore étudiée dans le diabète, à notre connaissance.

Technique utilisée

Les dosages ont été effectués par la méthode colorimétrique à l'hydroxylamine-chlorure ferrique, selon HESTRIN [10], titrant l'ester restant après incubation du sérum, à dilution convenable, avec le substrat: acétylcholine pour la cholinestérase, phénylacétate pour l'arylestérase, mais, dans ce cas, en présence d'ésérine, afin d'éliminer l'action de la cholinestérase sur ce substrat (technique de VINCENT, SEGONZAC et SESQUE [20]).[1]

Les résultats sont donnés en mg de substrat hydrolysé en 1 h, à 38°C, par 1 ml de sérum (mg/h/ml) et en unités enzymatiques internationales (μM/min/ml).

[1] Avec l'aide technique de Mesdames J. MAGRON et G. AUGOYARD-AUDIBERT.

Résultats

Nos résultats, portant sur 80 cas de diabète, sont répartis en 3 groupes principaux:

1. diabète I, non insulino-dépendant (diabète gras, diabète de l'âge mur);
2. diabète II, insulino-dépendant (diabète juvénile);
3. diabète associé, où la coexistence d'une autre maladie peut entraîner des modifications particulières des activités enzymatiques étudiées.

Un tel classement présente quelques difficultés dans certains cas de formes intermédiaires de passage au diabète métapléthorique, où s'installe secondairement une insuffisance insulinique.

Nous donnons ci-dessous, en trois tableaux, les résultats obtenus dans ces trois grands groupes d'observations.

Tableau I. Activités cholinestérasique et arylestérasique du sérum dans le diabète non insulino-dépendant (1er groupe, 41 cas)

		mg/h/ml	µM/min/ml	Ecart par rapport à la moyenne normale cholinestérase 23 arylestérase 280
Cholinestérase sérique	*Moyenne:* Extrêmes:	*32,9*±1,3 (19,2–50)	3,01±0,11 (1,75–4,57)	+43% p=0,001
Arylestérase sérique	*Moyenne:* Extrêmes:	*350*±14 (162–500)	42,7±1,7 (19,7)–61,0)	+25% p=0,02

Commentaires

Le *1er groupe, diabète I non insulino-dépendant*, donne des résultats appelant peu de commentaires: l'augmentation de la cholinestérase sérique (+43% en moyenne par rapport à la moyenne normale) est très significative. Ajoutons que le détail des résultats montre que la limite supérieure de la normale (28) est dépassée dans 72% des cas et qu'il n'en est aucun qui se trouve abaissé au-dessous de la limite inférieure de la normale (18). L'arylestérase augmente

Tableau II. Activités cholinestérasique et arylestérasique du sérum dans le diabète insulino-dépendant (2e groupe, 18 cas)

		mg/h/ml	μM/min/ml	Ecart par rapport à la moyenne normale cholinestérase 23 arylestérase 280
Cholinestérase sérique	*Moyenne:* Extrêmes:	*23,7*±2,0 (11,2–37,4)	2,16±0,18 (1,02–3,42)	+3% p=0,3
Arylestérase sérique	*Moyenne:* Extrêmes:	*329*±15,6 (230–430)	40,1±1,9 (28–52,4)	+17% p=0,05

Tableau III. Activités cholinestérasique et arylestérasique du sérum dans le diabète associé (3e groupe, 36 cas)

		mg/h/ml	μM/min/ml	Ecart par rapport à la moyenne normale cholinestérase 23 arylestérase 280
Cholinestérase sérique (20 cas)	*Moyenne:* Extrêmes:	*20,1*±8,9 (8,5–45.3)	1,83±0,80 (0,77–4,14)	–12%
Arylestérase sérique (16 cas)	*Moyenne:* Extrêmes:	*311*±90 (163–460)	97±27,4 (19,8–56,1)	+11%

parallèlement mais dans une moindre amplitude (+25% en moyenne), la limite supérieure de la normale n'étant dépassée que dans 40% des cas.

Beaucoup moins homogènes sont les résultats obtenus dans le *2e groupe (insulino-dépendant)* où l'activité de la cholinestérase, si l'on considère la moyenne, se situe très près de la moyenne normale. Ceci est dû à la présence, à peu près égale, dans ce groupe, de cas où la cholinestérase est augmentée (22% des cas), d'autres où elle est diminuée (27% des cas), les autres étant dans les limites normales. La cause de ces résultats disparates tient probablement à la plus grande gravité de cette forme de diabète et, de ce fait, à la fréquence relative de cas compliqués avec dénutrition et altération de l'état général, manifestations dont on sait qu'elles s'accompagnent d'une baisse plus ou moins considérable de la cholinestérase sérique. Par contre, dans ce groupe, le niveau d'activité de l'arylestérase est, en moyenne, encore augmenté, mais plus faiblement que dans le 1er groupe et aucun cas ne montre d'abaissement de cette activité enzymatique.

Il en est de même dans le *3e groupe, diabète associé*, où l'activité cholinestérasique est trouvée, en moyenne, diminuée, alors que l'arylestérase reste au contraire légèrement augmentée. Il semble donc que cette enzyme soit moins sensible aux perturbations pathologiques qui, associées au diabète, abaissent fréquemment la cholinestérase.

L'impact des lésions associées sur les activités estérasiques primitivement élevées du diabète est fort différent selon les cas. Ainsi les lésions vasculaires, qu'on peut considérer comme des processus évolutifs normaux du diabète, ne modifient que modérément ou nullement la cholinestérase sérique qui reste augmentée dans les cas compliqués d'artérite, même avec troubles trophiques (mal perforant plantaire), quand ils restent modérés. On sait d'ailleurs que l'athérosclérose s'accompagne fréquemment de l'augmentation de la cholinestérase. Par contre, l'apparition d'état cachectique ou infectieux grave entraîne la baisse de la cholinestérase: le résultat le plus bas parmi les cas étudiés concernait un diabétique grabataire avec quadriplégie, escarres et très mauvais état général. L'association avec les néoplasies a un effet analogue aboutissant à une baisse importante de l'activité cholinestérasique à une phase avancée de cancers retentissant sur l'état général, tels 3 cas de diabète associé à des néoplasmes du pancréas avec ictère où la chute cholinestérasique était très accentuée.

Enfin l'association avec la sclérose hépatique influence nettement la cholinestérase sérique des diabétiques: abaissement marqué dans un diabète avec cirrhose et dans un cas d'hémochromatose, tandis que des cas associés à une sclérose hépatique hypertrophique ou à une hépatite cholostatique donnaient des résultats subnormaux ou à la limite inférieure de la normale, alors qu'un diabète avec ictère cholédocien lithiasique gardait sa cholinestérase à un niveau élevé. En général, l'arylestérase varie dans le même sens, mais avec moins d'amplitude semble-t-il.

Quelques commentaires seraient utiles en ce qui concerne le *diabète métapléthorique*, mais ici le nombre réduit de cas étudiés (9 cas) ne permet pas de conclusion formelle. Nous avons obtenu, malgré une grande dispersion des résultats, une augmentation encore importante de la cholinestérase

(+ 35% par rapport à la moyenne normale), mais ceci est dû à quelques cas à cholinestérase très élevée, tandis que d'autres étaient notablement abaissés. L'arylestérase restait en moyenne aussi nettement augmentée dans ce groupe.

Aucune modification caractéristique n'a été observée en fonction de l'*acétonurie* (24 cas), Cholinestérase et arylestérase y sont, en moyenne, également augmentées (+ 25% par rapport à la moyenne normale), mais on note un nombre considérable de cas à activité diminuée. Il convient de remarquer que l'intensité de l'acétonurie était très variable et qu'il aurait été nécessaire de faire des dosages en série, sur des cas convenablement choisis, pour confirmer les observations de Kaswin [11] qui a signalé une très forte augmentation de la cholinestérase sérique, au cours des poussées acétonémiques du diabète.

Le comportement de la cholinestérase sérique dans le *coma* au cours du diabète ne peut être défini par nos résultats trop peu nombreux dans ce domaine (6 cas) et du fait de la diversité des comas chez les diabétiques. Notons dans les quelques cas étudiés, en moyenne, une légère diminution de la cholinestérase sérique (— 13%) alors que l'arylestérase était trouvée assez fortement augmentée (+ 34%). On ne saurait cependant affirmer, sur un nombre de cas aussi réduit, que le retentissement des processus physiopathologiques qui conditionnent le coma est plus marqué sur la cholinestérase, qui s'abaisse, que sur l'arylestérase qui reste élevée. D'ailleurs il convient de signaler que les prélèvements ont été effectués parfois après la phase aiguë comateuse, qu'il ne s'agissait pas toujours de coma acidocétosique classique, que l'hyperosmolarité pouvait y jouer un rôle important et que dans un de nos cas, il s'agissait d'un coma hypoglycémique. Aussi ne donnons-nous ces résultats que sous réserve, une étude plus large et plus précise étant indispensable dans ces cas souvent fort complexes.

Les *relations avec la lipidémie* sont intéressantes à considérer dans le diabète où l'hyperlipidémie est fréquente. Dans 29 de nos cas, nous avons effectué, parallèlement aux dosages de cholinestérase et d'arylestérase, la détermination de l'indice lipidique par la réaction au phénol de Kunkel et par la réaction de Burstein à l'héparine, dosant plus sélectivement les β-lipoprotéines. L'augmentation parallèle des tests lipidiques et des activités estérasiques a été trouvée dans 40 à 60% des cas, selon la réaction et l'enzyme considérées. La relation restait donc imparfaite, mais était plus fidèle avec la réaction de Kunkel qu'avec celle de Burstein. Une étude plus approfondie serait utile, notamment par comparaison avec le taux d'acides gras non estérifiés, pour préciser les rapports estérases-lipides dans le diabète.

La *relation avec l'insuline* est difficile à établir d'après nos résultats. On a signalé que la cholinestérase élevée des diabétiques graves non équilibrés s'abaisse relativement sous l'influence de l'insuline [PASSOUANT *et al.*, 13]. L'insuline, par elle-même, ne paraît pas avoir d'effet direct sur la cholinestérase: l'inhibition notée par HERSCHBERG [8] *in vitro* sur la cholinestérase du sérum de cheval, n'a pas été retrouvée *in vivo* sur le chien [BOVET et SANTENOISE, 3], ni chez le lapin par l'un de nous au cours de l'hypoglycémie insulinique [VINCENT, 17], ni chez l'homme, au cours du coma de la cure de Sackel dans la schizophrénie [VINCENT, 17, 18] où RANDALL et JELLINEK [15, 16] ont même trouvé son activité augmentée. Nos résultats dans le diabète II, insulino-dépendant, dont la plupart des cas se rapportaient à des sujets au cours d'équilibration de leur traitement insulinique, indiquent, par la moyenne observée, qu'une normalisation du niveau de la cholinestérase a pu être obtenue par l'insuline, en supposant que sans ce traitement la cholinestérase aurait été augmentée, ce que nous ne saurions affirmer. Si l'insuline produit un effet régulateur sur la cholinestérase sérique, c'est sans doute un effet indirect, à la faveur de modifications métaboliques sur les lipides, s'accompagnant d'une adaptation enzymatique des estérases, primitivement augmentées, à un métabolisme lipidique amélioré. La comparaison systématique avec le niveau d'insulinémie endogène, en l'absence de traitement insulinique, nous semble capable d'apporter, sur ce problème, des renseignements utiles que nous espérons obtenir dans la suite de ces recherches.

Conclusions

Cette étude nous permet les conclusions suivantes:

Dans le diabète I, non insulino-dépendant, diabète gras, l'activité de la cholinestérase sérique est, en général, fortement augmentée (72% des cas au-dessus de la limite supérieure de la normale). L'hypercholinestérasémie y est très significative.

Dans le diabète II, insulino-dépendant, cette augmentation d'activité est beaucoup moins fréquente et une baisse y est trouvée dans un nombre considérable de cas.

Il en est de même dans le diabète associé à des troubles pathologiques divers (tuberculose, sclérose hépatique, états infectieux, néoplasies avec mauvais état général) s'accompagnant d'une baisse de la cholinestérase sérique qui peut ainsi venir annuler l'hypercholinestérasie primitive du diabète.

L'arylestérase présente, en général avec moins d'amplitude, des variations semblables.

La relation avec la lipémie, réelle dans l'ensemble, apparaît cependant difficile à préciser.

Summary

Study on 80 cases of diabetes: serum cholinesterase increases heavily (72 per cent of cases) in the 'not-insulin-subjected' diabetes ('fatty diabetes'), much less frequently in the insulin-subjected diabetes, where its level often decreases considerably. Likewise in the associated diabetes (tuberculosis, hepatic sclerosis, neoplasias with bad general condition). Arylesterase shows similar variations, but less widely. The connection with lipemia, though actual, seems difficult to state accurately, if we compare with the results of Kunkel-phenol and Burnstein reactions.

Bibliographie

1. ALDRIDGE, W.: Serum esterases. I. Two types of esterases (A and B) hydrolysing p-nitrophenylacetate, propionate and butyrate, and a method for their determination. Biochem. J. *53:* 110–117 (1953).
2. AUGUSTINSSON, K.: Electrophoretic studies on blood plasma esterases. Acta chem. scand. *13:* 571–592 (1959).
3. BOVET, D. et SANTENOISE, D.: Action de la vagotonine sur le pouvoir cholinestérasique du sérum sanguin. C. R. Soc. Biol. *135:* 844–847 (1941).
4. CRISTOL, P.; PASSOUANT, P.; BENEZECH, C. et DUTARTRE, J.: Valeur séméiologique de la cholinestérase sérique en clinique. Presse méd. *54:* 557–558 (1946).
5. DHELLEMMES, G.: Recherches sur la pseudocholinestérase du sérum humain; Thèse Pharmacie Lille (1947).
6. DROUET, P.; VERAIN, M. et FRANQUIN, A.: Recherches sur la cholinestérase du sérum humain à l'état normal et pathologique. Bull. Acad. Méd. *126:* 412–414 (1942).
7. DUCARRE, M.: La pseudocholinestérase sérique; Thèse Médecine Lyon (1968).
8. FABER, M.: Serum cholinesterase in deseases. Acta med. scand. *114:* 60–71 (1943).
9. HERSCHBERG, A.: Contribution à l'étude de la régulation physiologique du système acétylcholine-cholinestérase; Thèse Sciences Paris (1946).
10 HESTRIN, S.: The reaction of acetylcholine and other carboxylic derivatives with hydroxylamin and its analytical applications. J. biol. Chem. *180:* 249 (1959).
11. KASWIN, A.: La cholinestérase du sérum humain à l'état physiologique et dans divers états pathologiques. Presse méd. *53:* 713–714 (1945).
12. MOUNTER, L. and WHITTAKER, V.: The hydrolysis of esters of phenol by cholinesterase and other esterases. Biochem. J. *54:* 551–559 (1953).
13. PASSOUANT, P.; BENEZECH, C. et DUTARTRE, J.: Considérations sur la valeur de la cholinestérase sérique en clinique. Presse méd. *53:* 223 (1945).
14. PRAT, J. de: La cholinestérase du sérum. Application clinique; Thèse Pharmacie Toulouse (1945).

15. RANDALL, L.: The effect of insulin on serum lipids and cholinesterase in schizophrenia. Lab. clin. Med. *25:* 1025–1028 (1940).
16. RANDALL, L. and JELLINEK, E.: Physiological studies on insulin treatment of acute schizophrenia. Endocrinology *25:* 278–281 (1939).
17. VINCENT, D.: Cholinestérase sérique et hypoglycémie insulinique C. R. Soc. Biol. *142:* 1137–1139 (1948).
18. VINCENT, D. et PARANT, M.: L'activation des estérases sériques dans le diabète. Toulouse méd. *52:* 775–776 (1951).
19. VINCENT, D. et SEGONZAC, G.: Cholinestérase sérique et réaction au phénol de Kunkel. Rev. franc. Et. clin. biol. *3:* 163–166 (1958).
20. VINCENT, D.; SEGONZAG, G. et SESQUÉ, G.: Technique de dosage des estérases A (arylestérase) et C (cholinestérase) du sérum humain. Ann. Biol. Clin. *18:* 489–492 (1960).

Adresse des auteurs: Pr D. VINCENT, Laboratoire de Biochimie appliquée, Faculté de Médecine, 8, avenue Rockefeller, *69-Lyon 8e;* Pr. M. PLAUCHU, Hôtel-Dieu, place de l'Hôpital, *69-Lyon 2e;* Pr M. GIRARD, Hôpital de la Croix-Rousse, Grande-Rue de la Croix-Rousse, *69-Lyon 4e* (France)

Authors' Index

6th International Congress of Clinical Chemistry

Munich (Germany), July 26–30, 1966

Edited by O. WIELAND, Munich

Vol. 1: Clinical Protein Chemistry

Main Lectures

Abstracts of Short Lectures

Vol. 2: Clinical Enzymology

Biochemistry and Diagnostic Value of Isoenzymes

Immunochemical Enzyme-Studies

Diagnostic Value of Enzyme Determination

Methods of Enzyme Determination

Vol. 3: Progress in Clinico-Chemical Methods

Main Lectures

Abstracts of Short Lectures

Vol. 4: Advances in Clinico-Biochemical Research

Main Lectures

Abstracts of Short Lectures